Community
Organizing

This book is dedicated

to the volunteers in this program,
and the thousands like them
in communities across the country,
who are comitting themselves
to make the nation's cities better places to live,

and to our families.

Community Organizing

Building Social Capital as a Development Strategy

Ross Gittell • Avis Vidal

SAGE Publications
International Educational and Professional Publisher
Thousand Oaks London New Delhi

For information:

SAGE Publications, Inc.
2455 Teller Road
Thousand Oaks, California 91320
E-mail: order@sagepub.com

SAGE Publications Ltd.
6 Bonhill Street
London EC2A 4PU
United Kingdom

SAGE Publications India Pvt. Ltd.
M-32 Market
Greater Kailash I
New Delhi 110 048 India

Printed in the United States of America

Library of Congress Cataloging-in-Publication Data

Gittell, Ross J., 1957-
    Community organizing: Building social capital as a development
strategy / by Ross Gittell and Avis Vidal.
        p. cm.
    Includes bibliographical references and index.
    ISBN 0-8039-5791-2 (cloth: acid-free paper). — ISBN
0-8039-5792-0 (pbk.: acid-free paper)
        1. Community development—United States—Case studies.
    2. Community development corporations—United States—Case studies.
    3. Community organization—United States—Case studies.   I. Vidal,
Avis. II. Title.
    HN90.C6G57 1998
    307.1'4'0973—dc21   97-45463

    99 00 01 02 03 10 9 8 7 6 5 4 3 2

Acquiring Editor:	Catherine Rossbach
Editorial Assistant:	Kathleen Derby
Production Editor:	Diana E. Axelsen
Editorial Assistant:	Denise Santoyo
Typesetter/Designer:	Danielle Dillahunt
Indexer:	Virgil Diodato
Cover Designer:	Candice Harman

Contents

Preface

Many program evaluations—especially those (like this one) that rely heavily on program participants for their information, and that are supported financially by funders who also supported the intervention under study and who are publicly identified with it—are mainly descriptive and focus almost exclusively on positive outcomes. This rendering of one initiative's experience is not. It is both detailed and analytical, because we believe that getting inside the workings of an innovative program and understanding the thinking that guides the actions of the participants provides rich learning opportunities. We have tried to capitalize on those opportunities in our analysis. In the process, we have tried to depict the program and its participants fairly and honestly and to help readers learn from their failures as well as their successes.

The examination of experience in the three sites, which forms the core of the book, illustrates the complexity and difficulty of community development. Although the basic description of the Local Initiatives Support Corporation (LISC) demonstration program often appears straightforward, program implementation is multifaceted, strategic, and sometimes even artful. It clearly requires hard work and high levels of commitment on the part of staff, members of the support community, and resident volunteers.

That there were some setbacks, and even some failures, is disappointing—sometimes even painful, most especially to those most directly affected by them—but it is not surprising. Community development is challenging, pains-

ability to secure both funding for our research and candid information from program participants. We would also like to thank Robyne Turner, whose detailed observations of the program in Palm Beach County helped deepen our understanding of the program's complex dynamics at the neighborhood level.

Funding for our research was provided by the John D. and Catherine T. MacArthur Foundation and the Charles Stewart Mott Foundation. We are grateful not only for their financial support but also for their ongoing interest. We also appreciate the personal encouragement of interested individuals at these organizations: Rebecca Riley, Bonnie Weaver, and Ralph Hamilton at MacArthur, and Ruth Goins and Jack Litzenberg at Mott.

Our editors at Sage, Carrie Mullen and Catherine Rossbach, have been enthusiastic supporters of our work. We thank them both for their energy and patience, and thank Catherine for her help as we put together the final manuscript for publication.

Finally, we each enjoyed the benefit of more personal support. For Ross, Jody Hoffer Gittell and Marilyn Gittell provided ongoing intellectual guidance for the work. Jody, together with Rose Hoffer Gittell, provided daily enthusiastic encouragement. Marilyn and Irwin Gittell provoked intellectual curiosity and social concern, and also provided strong personal support. For Avis, the ups and downs during the project have been smoothed and softened by the interest, supportiveness, and kindness of Prue Brown, Karen Courtney, Dennis Derryck, Julie Freiesen, Anne Kubisch, Jean MacMillan, Stephanie Nickerson, Harold Richman, Alex Schwartz, and Nick Weiner. Of special value have been the warmth and affection of Barbara Pizer and, until his death, Sam. Thank you all.

Introduction

The community development movement has now matured into what many would call a fledgling industry that includes more than 2,000 community development corporations (CDCs; National Congress for Community Economic Development, 1995). The community development field includes an increasingly sophisticated network of foundations, corporations, intermediaries, and technical assistance providers, with local, state, and federal government agencies also playing key roles in localities where CDCs are numerous (Keyes et al. 1996).

The speed of changes in the funding landscape and deteriorating conditions in America's inner cities (Jargowsky 1996; Mills and Lubuele 1997; Wilson 1996) lend urgency to the need to broaden and deepen community development capacity locally and nationally. The devolution of responsibility to the state and local levels will increase the demands made on the community development field, as will the inevitable political battles for diminishing resources. These battles, formerly fought primarily in Washington, will now be waged locally. Neighborhood residents, CDCs (and other organized groups), and the national and local community development support community (e.g., private foundations, community development intermediaries, and advocacy organizations) will need to play an active role to address the great needs of low-income neighborhoods and communities of color.

area support community. These commitments are particularly impressive given the limited physical and economic improvement in the Valley to date.[4]

Eichler felt that consensus organizing held promise as a vehicle for generating citizen action in a variety of environments, but he needed an organizational base from which to work. The effectiveness of using the core LISC program to mobilize both community members and supporters was clear both to him and to LISC.[5] A partnership was born.

Core Strategies

Modeled on efforts in the Mon Valley, the LISC demonstration attempted to achieve its main objectives through core strategies and activities, applied similarly in all three sites. The *core strategies* included broad involvement based on consensus and the promise of delivering tangible products (housing production); deconstruction of complex undertakings into a series of relatively simple, straightforward activities; the engagement of carefully selected targeted neighborhoods and support context individuals; leadership development (e.g., in neighborhoods on CDC boards and as advocates for community development in the support community); parallel organizing of residents in multiple neighborhoods and members of the support community, using progress in each of the individual "spheres" to foster effort and progress in the other; linking residents of low-income neighborhoods to resources and individuals in the support community; and risk management (e.g., selecting sites and neighborhoods with a good chance of success, yet not avoiding tough challenges).

Program Activities:
The "Nuts and Bolts" of Implementation

The LISC demonstration approach involved a sequence of coordinated and interrelated activities.[6] Although there was a "model" program design fielded in all sites, some alterations and variations in implementation of the model occurred over time and across sites. Some of the variation was related to differences in local contexts and circumstances. Other variations reflected formative cross-site learning, that is, alterations suggested by Eichler and local coordinators based on the experience in other sites. Still other variations related to different views among the coordinators about what would work best.

A core set of activities, in sequence, was implemented in all of the sites. The first activity was *site assessment.* Site assessment involved LISC staff determining the feasibility of operating a successful demonstration effort. As with other potential LISC sites, assessment included extensive interviews with potential funders and program supporters to determine if the local match money LISC required could be raised and whether more general support for LISC efforts would be forthcoming. Distinctive from other LISC prospective sites, the assessment also included evaluation of the potential to organize new community groups using the consensus organizing approach.

After an area was selected as a demonstration site, the program began fund-raising and hired "development team" staff: a local coordinator and three community organizers. The development team's main responsibility was to recruit neighborhood volunteers and work directly with them to enhance their capacity—that is, their organizational, political, and technical capacity—to affect their community's development. In addition, local area professionals were recruited to serve as technical consultants to the neighborhood groups.[7] Local coordinators were responsible for hiring, training, and managing community organizers and technical consultants.

One of the development team's first responsibilities in each site was the selection of six neighborhoods in which to target its efforts.[8] Each community organizer identified potential volunteers in two assigned neighborhoods and assisted them in establishing CDCs that had broad-based representation. Each new CDC held an open town meeting to discuss community priorities. The town meeting, along with technical training and assistance, was meant to help board members identify projects that had potential to benefit the community as a whole. Then, by undertaking real estate development projects under the guidance of development team members and with LISC and private and public-sector funding, CDC volunteer board members were expected to further their commitment to their community's development, learn about the development process, expand their organization's capacity and establish its credibility, and gain control over their community's development.

Each site had a planned date to *transition* (initially two years in Little Rock and Palm Beach County and three years in New Orleans) from a development team managed by Michael Eichler to another arrangement. However, at program commencement nationally and at the individual sites, the exact form that transition would take—namely, institutional structure, staffing, governance, funding—was less clear than the transition date itself.

Demonstration Program Management

Michael Eichler (national LISC program director) and Richard Manson (vice president of LISC) had main management responsibilities for the LISC demonstration (see Figure 1.1). Both were employees of LISC and reported to LISC Senior Vice President Andy Ditton. Eichler and Manson shared responsibility for site assessment and local fund-raising. In addition, Eichler had primary responsibility for hiring and supervising local development teams, and working with the local coordinators on strategy. Richard Manson was responsible for establishing LISC Local Advisory Committees, working with CDCs and their

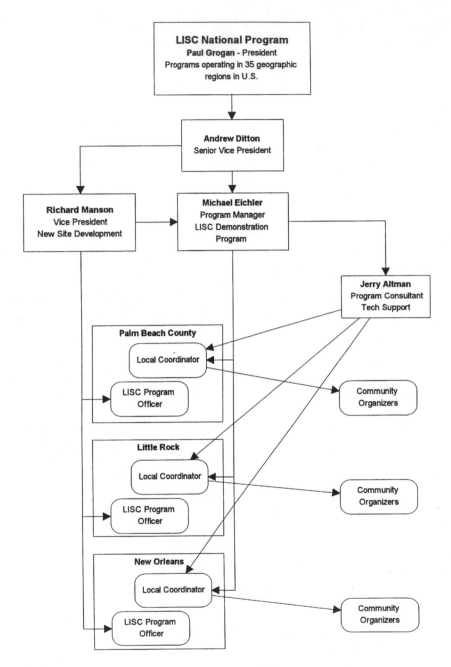

Figure 1.1. LISC National Demonstration Program Organization

working committees on their first projects, hiring and supervising a local LISC program director, and reviewing Eichler's performance on an annual basis.

Michael Eichler and the development teams were assisted in all the sites by a national consultant with legal training and extensive community development experience, Jerry Altman. Altman was primarily responsible for (1) technical training of development team staff (including technical consultants) and CDC board members in such topics as finance, marketing, and project development; and (2) project development support. Altman employed a community development project-development model that included CDCs' identifying target areas, forming working committees (e.g., marketing and counseling, finance) and undertaking technical analysis.

Michael Eichler trained each local coordinator in a series of intensive one-on-one sessions. The community organizers received both formal and informal training. The formal training sessions lasted two weeks and were conducted by Michael Eichler, the local coordinator, and Jerry Altman. Informal training was ongoing and interactive.

Eichler managed by leadership and inspiration. He had high expectations of all coordinators and required that they share core values (i.e., commitment to the targeted population and the enhancement of their capacities and control) and work toward a common mission (i.e., the realization of the primary objectives of the national demonstration). Local coordinators were required to follow the sequenced activities outlined earlier; however, they had significant discretion over tactics used. Eichler assisted the coordinators whenever they requested it, but he rarely imposed his opinion on local coordinators, even if he doubted that their actions would be efficacious. His rationale was that it was important for coordinators to retain full responsibility for what happened in their sites, even if it meant they made (and hopefully learned from) mistakes.

As the sequence of program implementation activities was being undertaken in the targeted neighborhoods, Eichler, Manson, local development team staff, the local LISC program directors, and community groups were engaged in building relationships with the local support community.[9] The basis of the relationship building with the private sector was most unique as it was focused on the pursuit and realization of mutual gains, not simply goodwill. A gradual and evolutionary process of building understanding and working relationships was required because the sites were places where community-based development was still relatively new and underappreciated.

One important event, unanticipated when the program was designed and inaugurated in the field, both reflected and furthered differences in priorities among the major program actors. Michael Eichler left LISC in 1994 to establish the Consensus Organizing Institute (COI), an independent nonprofit organization dedicated to using consensus organizing to help residents of poor communities gain greater influence over changes affecting their neighborhoods. Under

Eichler's direction, COI's organizing facilitates resident action in a variety of arenas (e.g., education, safety, and employment in addition to physical and economic development), often working with one or more other entities, such as LISC, that have the capacity to deliver technical and financial assistance. LISC became COI's first client, with COI reporting to Richard Manson, and the local coordinators in the LISC demonstration sites became COI staff members; this changed the relationship between LISC and the development teams to a contractual one. Both the factors leading to this separation and the changed nature of the relationship had unfortunate program implications; these are discussed in detail in Chapters 7 and 8.

Learning From the
LISC National Demonstration

A central objective of this book is to distill lessons for the community development field from this timely initiative. Although the LISC demonstration is only one of a myriad of community development initiatives launched over the past decade, it is one of the most instructive, both because of its clearly structured design and because it was monitored by two outside observers (the authors) from start to finish.

Our monitoring included a mix of field visits, telephone interviews, and review of program-related materials.[10] Each of the sites was visited at least annually by one of the authors. During these field visits, we (1) interviewed major program participants, that is, development team and LISC staff, resident volunteers (particularly CDC officers), and members of the support community; and (2) observed program activities, including staff training sessions and staff meetings, CDC activities in targeted neighborhoods (e.g., town meetings and CDC and committee meetings), LISC Local Advisory Committee meetings, special events, and meetings among officers of CDCs.[11] In addition, we conducted bimonthly telephone interviews with the local coordinators (and after transition, monthly ones with LISC program officers), monthly telephone interviews with Eichler, and semiannual conversations with Richard Manson. These contacts were supplemented by quarterly "all coordinator" conference calls in which the coordinators rotated picking topics to discuss.[12] At the end of the fieldwork (summer of 1996), we reviewed our preliminary findings with Eichler and Manson (in half-day "in-person" meetings) and with the coordinators and LISC program officers (in conference calls) and received comments and suggestions. Finally, we gave key participants, including the evaluation's funders,[13] an opportunity to read and comment on a draft of our report.

The findings and lessons we report are drawn primarily from a comparative process analysis of major program activities in the three demonstration sites, although we also make some comparisons to the pilot site (the Mon Valley) and other community development efforts. The lessons reflect not only an assessment of the strengths, weaknesses, and accomplishments of the LISC demonstration program, but also our thinking on some of the larger issues facing the community development field. These include the following:

- the potential of outside agents (i.e., a program intervention) to "jump-start" community revitalization efforts where they are currently lacking;
- the roles of those outside agents, including guidelines for community development partnerships and program management;
- strategies to broadly engage the private corporate sector in community development;
- the value and limits of a consensus-based approach to community organizing;
- the benefits and problems of focusing community efforts on real estate production;
- the key requisites for sustaining community development efforts; and
- the role of social capital in community development.

Toward this larger end, we divide the remainder of this volume into three parts. The first, consisting of Chapters 2 and 3, places the demonstration program in context, first conceptually, then programmatically. Chapter 2 draws on social capital and network theories to accomplish two things: (1) positioning community development within a broad conceptual framework and describing it in the vocabulary of those bodies of literature and (2) articulating an analytic framework appropriate specifically to community development interventions, and relating it to the broader discussion of social capital. We use this latter framework to describe the operating principles and program logic of the LISC demonstration.

Chapter 3 positions the LISC demonstration in the context of current programmatic developments in the community development field nationally. These include the rapid expansion of the CDC movement since 1980, foundation-funded comprehensive community change initiatives, the federal Empowerment Zone and Enterprise Community Program, and community organizing. Framed in social capital and network terms, these approaches to addressing the challenges of community development put the LISC demonstration in practical perspective, particularly for the purpose of setting appropriate performance and outcome parameters.

The analytic core of the book, Chapters 4 through 7, discusses the LISC program experience in detail, including both how activities in each site contributed to intermediate outcomes and how they in turn affected the potential for sustained activity and progress. We focus on four major sets of activities: (1) program start-up; (2) organizing and supporting the CDCs, including developing neighborhood leadership; (3) relationship building with the private sector; and (4) transition. Each activity analysis is divided into four segments: overall objectives and strategy, activities used to implement the strategy, a comparative analysis of the main findings from program experience, and lessons for the community development field.

The final section, Chapter 8, distills broader program and policy lessons for the community development field. It draws on the analysis in Chapters 4 to 7 to develop a set of key contributors to successful community development interventions and uses these to reflect on what the intervention has to teach us about the role of social capital in community development. Our focus in Chapter 8 is on how social capital can (and cannot) be nurtured, on how social capital may contribute to community development, and on indicators of whether bonds and bridges fostered through consensus-based organizing are likely to be sustained and beneficial to the residents of the targeted low-income communities.

Notes

1. A survey conducted in 1989 of 133 U.S. cities with populations of more than 100,000 (75 percent of all cities) found one or more CDCs operating in 95 percent of them (Goetz 1993:117).

2. Since its inception in 1979, LISC has raised more than $2 billion in donations, investments, and below-market loans for CDCs, roughly 97 percent of it from the private sector. In 1995, LISC provided more than $450 million in grants, low-interest loans, and equity financing for inner-city revitalization projects and provided service and support to more than 1,400 CDCs nationwide (LISC, 1996). In the metropolitan areas in which it is active, LISC raises a pool of local funds, primarily from corporate and foundation sponsors, which are matched with funds from sources nationwide. LISC's success hinges heavily on the existence of established CDCs that have developed particular capabilities.

3. The Allegheny Conference is a private nonprofit organization founded in the early 1940s in response to Pittsburgh's deteriorating social and economic conditions. It has been credited with being a major contributor to the city of Pittsburgh's much-heralded Renaissance. Its executive committee comprises 25 chairmen of major Pittsburgh corporations as well as the president of the University of Pittsburgh.

4. For more detailed discussion of community development efforts in the Mon Valley, see R. Gittell (1992).

5. After his efforts in the Mon Valley, Eichler worked in Houston with LISC. Although there has been no systematic study of activities in Houston, there are indications that some progress in creating new groups and gaining the support of the private sector has been made (United Way, 1996). However, most indicators are that efforts in Houston were not as successful as in the Mon Valley. In Houston, it proved more difficult to organize residents and retain resident interest and commitment to CDCs.

6. The approach has many complexities and requires many decisions within the context of ongoing program activity; however, its basic operation can be represented fairly simply. In fact, one of the approach's key benefits is the ease with which it can be described to, and understood by, those with limited community development experience, including members of the metropolitan area support community targeted by the demonstration program.

7. Technical consultants came with different professional expertise, including law, architecture, and marketing, and worked on a part-time basis. They were assigned to work with individual neighborhood groups and community organizers by the coordinator.

8. In New Orleans, seven neighborhoods were initially targeted. Eventually efforts in one neighborhood, Gert Town, were discontinued.

9. What makes LISC and the LISC demonstration most distinctive is the focus on relationship building with the private sector—most community development initiatives give priority to relationships (often with the focus on securing money) with the public sector.

10. In all three demonstration sites, the authors tried to establish working relationships with scholars at local institutions to help document local program activities. This was intended not only to enhance the amount and quality of information collected, but also to help build local research capacity on community development. Only in Palm Beach County were the authors able to engage a capable and committed local researcher, Robyne Turner, Associate Professor at Florida Atlantic University.

11. The authors spent a significant portion of their time in the field with community organizers visiting targeted neighborhoods, meeting with community residents, and attending program-related activities.

12. In these quarterly sessions, the principal investigators initially acted mostly as facilitators; but, over time, these calls and the research itself became more formative in character and appeared to have some influence on implementation.

13. The evaluation was supported financially by the John D. and Catherine T. MacArthur Foundation and the Charles Stewart Mott Foundation.

CHAPTER 2

Social Capital and Networks in Community Development

Framing the LISC Demonstration

There is increasing recognition of the potential role of social capital and networks in community development, both for understanding it conceptually and for strengthening practice. For example, Keyes et al. (1996) suggest that national community development intermediaries (e.g., LISC) are crucial in providing nonprofits located in institutionally barren environments with the necessary financial, technical, political, and moral support needed to foster a viable low-income housing industry. They further suggest that it is incumbent on the intermediaries, with their strong corporate and foundation backing, to cultivate supportive local networks for nonprofit housing groups (p. 26). Hornburg and Lang (1998) conjecture that building social capital may give people and communities the connectedness they need to face the new realities of devolution. Yet, there remains a paucity of detailed empirical effort devoted to testing and refining these ideas and those reviewed subsequently. The LISC demonstration provides us with a valuable opportunity to test some of these views.

Social Capital and
Community Development

Since Robert Putnam's identification of the role of social capital in regional governance and economic development in Italy (1993) and his later suggestion of its importance in the United States (1995a, 1995b), there has been a virtual industry of interest and action created around the implications of his findings for the development of low-income communities.[1] Although there has been other literature on social capital and related topics—including Coleman (1988), Jacobs (1961), and Wilson (1987)—to draw on, Putnam's work has often been the central motivation as well as the intellectual departure point.

The industry of interest in social capital has taken the form of an effort by scholars to define *social capital* as it relates to the development of low-income inner cities in the United States. It has incorporated the relatively recent focus in the community development field on "civic capacity" and "community building" *and* reinforced the long-standing preference of practitioners and prominent national foundations for framing their community development activities in asset-based terms, that is, in terms of "the capacity of communities to act" rather than of "need."

We begin this chapter with a review of the emerging literature defining social capital, supplementing that discussion by introducing related concepts from network theory. The main objective is to provide a rich contextual understanding of the LISC demonstration and the central elements of its program design. Then, in the second part of the chapter, we present a more detailed framework that incorporates the main elements of the analytical constructs reviewed, but tailors them to the more concrete task of evaluating a community development intervention. We use both in subsequent chapters to characterize major streams of contemporary community development practice (Chapter 3) and to analyze the LISC demonstration (Chapters 4-7).

Putnam and His "Followers"

Robert Putnam, in work applying lessons from *Making Democracy Work* to the U.S. experience, tries to address the question of why there is such a wide variation in the well-being of our nation's communities and why there has been a general decline in civic engagement in the United States over time.[2] He asked questions similar to those asked with increased frequency by federal government agencies and foundation grant makers when they observe the widely divergent impact that the same level of funding and the same program have on different cities and on different neighborhoods in the same city.

Putnam attributes a significant portion of differences in government effectiveness, economic health, and community well-being to the presence (versus absence) of social capital. Stated simply, the main elements of social capital for Putnam are trust and cooperation.[3] Social capital consists of networks and norms that enable participants to act together effectively to pursue shared objectives.[4] Again at the risk of oversimplifying, there are two main types of social capital according to Putnam—the type that brings closer together people who already know each other (we call this *bonding capital*), and the type that brings together people or groups who previously did not know each other (Putnam called this *bridging capital* and we adopt his term). Putnam's theory of social capital presumes that the more people connect with each other, the more they will trust each other, and the better off they are individually and collectively, because there is a strong collective aspect to social capital: The social and economic system as a whole functions better because of the ties among actors that make it up (Briggs 1998).

Although Putnam's work has been subjected to considerable critique,[5] especially as it may relate to the development of low-income communities in the United States, it does have value in helping to frame and better understand community development practice, including some of the objectives and strategies of the LISC demonstration. In many respects, the LISC demonstration tried to enhance social capital, building both bonds and bridges within low-income communities and fostering bridges between residents of low-income communities and the larger metropolitan area support community.

Within low-income communities, the LISC demonstration sought to establish CDCs that crossed racial, ethnic, and class lines and brought residents together with business owners and managers from local nonprofit organizations such as hospitals and social service agencies. Across low-income communities, new linkages were attempted through coalitions of CDCs. In the metropolitan area, the program tried to strengthen bonds among members of private-sector and philanthropic organizations who already knew each other from other settings. Finally, if community residents and institutions were strengthened in these ways, they would have the opportunity to establish new bridges to outside resources also organized by LISC. In fact, a core aspect of the LISC demonstration was the fostering of bridges between the private sector in the larger metropolitan area and the newly organized CDCs. These new relations were intended to go beyond providing investment capital and charitable contributions to include technical and political support for the CDCs' efforts.[6]

Bridging social capital is, in fact, a key element of the LISC demonstration. Absent its focus on bridging, the character of the social capital construction attempted by LISC would have been much different. First, without the cultivation of bridges internal to targeted communities, the new CDCs could "degenerate into a congeries of rent-seeking special interests" (Foley and Edwards 1996).[7] Second, given the nature of the neighborhoods targeted, without links to outside resources and opportunities, the stronger ties developed internal to

Authors	Key Elements of Social Capital	Additions to Putnam
Putnam, 1993, 1995a, 1995b	Trust, cooperation, long-term relationships	
Temkin & Rohe, 1997	Sociocultural milieu and institutional infrastructure	Neighborhood ability to act on common interest
Briggs, 1997	Social capital as leverage and social support	Social capital as leverage
Keyes, Schwartz, Vidal, & Bratt, 1996	Long-term trust and relationships, shared vision, economic incentives to mutual interest, financial nexus	Networking, shared vision, financial nexus
Powell, 1990		See Keyes et al.
Granovetter, 1993, 1994		The strength of weak ties
Burt, 1992		Structural holes

Figure 2.1. Key Elements of Social Capital

the communities (especially the CDC) might not be efficacious because they would lack links to outside resources and opportunities. This would make it difficult to motivate and sustain community development activity.

There are many variations—some more or less subtle and some more important than others—regarding how to best define social capital as related to community development practice (see Figure 2.1). For example, the Committee for Economic Development (1995:12-13) in *Rebuilding Inner City Communities* defines social capital as the resources embedded in social relations among persons and organizations that facilitate cooperation and collaboration in communities. This definition of social capital comes close to Putnam's, but does not make the important distinction between bonding and bridging capital. Most other commentators distinguish between the two, and they are relevant to our discussion of the LISC demonstration.

Perhaps the most noteworthy, and certainly the most empirically grounded, effort to test the efficacy of social capital in the U.S. community development context is by Temkin and Rohe (1998).[8] Their focus is on the role social capital plays in shaping community change. They use neighborhoods in Pittsburgh, Pennsylvania, to model this change and provide an empirical baseline against which social capital can be measured and tested.

Temkin and Rohe make significant advances in refining Putnam's concept of social capital. They describe social capital as consisting of two main components: sociocultural milieu and institutional infrastructure. Sociocultural milieu is quite similar to the bonding capital outlined by Putnam. Institutional infrastructure has strong similarities to bridging capital.

Temkin and Rohe (1998) operationalize sociocultural milieu to include the degree to which residents feel their neighborhood is a spatially distinct place; interact with one another in the form of borrowing small items, visiting, discussing local problems, and helping each other with small tasks; work and socialize in the neighborhood; and use neighborhood facilities for worship and grocery shopping (p. 13).[9] Their definition of institutional infrastructure includes variables measuring the presence and quality of neighborhood organizations, voting by residents, volunteer efforts and the degree to which those are focused on neighborhood issues, and the visibility of the neighborhood to citywide officials—as measured by whether or not the neighborhood is in an area covered by a well-funded CDC, as well as the presence of large institutions in the area (p. 19). Institutional infrastructure measures the level and quality of the organizational ability of neighborhoods to act on their common interest. The institutional infrastructure relates not only to the presence of community groups but also to the existence of communication between the neighborhood and the larger city. The latter point is why we relate Temkin and Rohe's institutional infrastructure to bridging capital. However, it is also important to recognize the internal organizational capacity component of their definition. This was a critical element of the LISC demonstration: to build neighborhood-based organizational capacity in the form of CDCs.

For Temkin and Rohe, together these two—sociocultural milieu and institutional infrastructure—help measure the degree to which a neighborhood has residents who are committed to a spatially circumscribed part of a city and have the wherewithal to turn this commitment into effective collective action. As described by Temkin and Rohe, and as we shall argue with the LISC demonstration, the effective blending of the two (for Temkin and Rohe, sociocultural milieu and institutional infrastructure; for Putnam and us, bonding and bridging capital) is critical.

Temkin and Rohe conclude that social capital matters—that it is a key determinant in predicting neighborhood stability and that neighborhoods with relatively large amounts of social capital are less likely to decline when other factors are held constant (p. 26). Their empirical work identifies two key components of social capital affecting neighborhoods: the overall sense of attachment and loyalty among neighborhood residents and the ability of residents to leverage a strong sociocultural milieu into effective collective action.[10] The former could be thought of as the commitment of residents to each other and their environs, and the latter their capacity to act. The latter could include individual (e.g., leadership) and organizational (e.g., CDC) capacities. The enhancement of resident commitment and capacity were two key objectives of the LISC demonstration. The former could increase bonds and the latter the potential to bridge, that is, to make external connections that would leverage external resources and support.

ded in a racially diverse network in LA has a significant positive effect on the incomes of both black and Hispanic men and women. There was a $24,213 difference in incomes of those with a racial bridge and those who did not have the benefit of such network ties" (p. 5).

The strength of weak ties also seems relevant to the LISC demonstration. One of the key deficiencies in many low-income communities is the lack of linkage to the larger metropolitan area opportunity structure, including financial, technical, social, and political resources. Following Granovetter's logic suggests the efficacy of community development efforts, such as the LISC demonstration, that help establish a network of weak ties to organizations and individuals outside the inner city from whom individuals and organized groups in the inner city could garner resources (e.g., financial, personal, and professional expertise) and political support.

The LISC demonstration attempted to do this by the parallel organizing of community residents in low-income neighborhoods and a metropolitan community development support network, and the linking of the two. The new weak ties between community groups (CDCs) and the support community could bring new ideas, resources, and opportunities to metro-area community development efforts. The establishment of new ties among people in a community could lead to new norms of trust and cooperation and eventually to new activities and collective action that could be beneficial for the community as a whole (which can be related back to the inclusion by Keyes et al. [1996], of mutual gains in social capital). In the LISC demonstration program, the fostering of new internal community ties would be achieved mainly through the organization and work of CDCs (focused on real estate development, but supplemented by smaller projects such as neighborhood cleanups, crime watches, and arts or cultural activities) designed to bring people together in new ways, to experience some success, and to build momentum, confidence, and stronger ties over time.

Ronald Burt (1992) adds useful sociological constructs to Granovetter and the discussion of social capital and community development. Burt's focus is on what he calls *structural holes*. Structural holes are similar to the social capital that requires bridging in Putnam's vernacular and "yet-to-be-made" weak ties in Granovetter's frame. They are defined as the "gaps between nonredundant contacts."[11] Stated more simply for our purposes here, structural holes are individuals not benefiting from connecting with others and with resources that could be beneficial.

In contrast to Putnam's presentation of social capital, in Burt's analysis the benefits from spanning structural holes are often not shared, but expropriated by "entrepreneurs." For example, there is evidence that in many cities, some groups and local politicians have functioned as entrepreneurs and captured (political) rents by taking advantage of the divisions among local residents and that many groups and individuals have an interest in the retention of divisions, not their elimination.[12] This logic suggests that what may be required to foster community development in areas that have lacked formative efforts are new associations and entrepreneurs with a broader social interest than self-interested groups and individuals. These entrepreneurs would not try to use social division to their own advantage, but instead attempt to fill structural holes to produce broader social benefits.

Burt suggests the value of processes at the local level that fill structural holes to serve a collective good. The challenge is to (1) find the appropriate people and organizations to fill structural holes and (2) counteract the incentive for some to sustain structural holes and fill them for their own benefit.

Burt's structural hole theory provides a new perspective on the LISC demonstration. LISC managers and the development team members could be viewed as community development entrepreneurs, attempting to produce benefits, not for themselves but for the residents of low-income neighborhoods in targeted metropolitan areas. They are meant to accomplish this by helping to fill structural holes within communities and between communities and outside resources with the organization of new CDCs and the linking of CDCs with a newly organized support community.[13] In the program design, community organizing, the formation of CDCs, and the establishment of a local LISC program are used to connect people within communities in new ways and to establish new connections between the residents of low-income communities and outside resources. In this way, LISC and the development teams bring new connections, new resources, and new capabilities to targeted communities by filling structural holes and spanning new sociocultural bridges.

What is suggested by our brief review of arguments by Putnam, Temkin and Rohe, Briggs, Keyes et al., Granovetter, and Burt is that perhaps the greatest potential value of the LISC demonstration was its effort to strengthen the foundation on which community development depends by instituting processes to increase social capital bonds and bridges and fill structural holes in community political and social space to the benefit of local residents.

Limits to Social Capital Construction and the LISC Demonstration

Increasing social capital where it is currently lacking is a challenging undertaking. This is particularly true in some of the low-income neighborhoods targeted by the LISC initiative. The targeted areas have suffered from years of decline and neglect. In many of these neighborhoods, the most successful and competent individuals and businesses move out when they can (Bates 1994; Wilson 1987), often leaving social and economic vacuums. In addition, these neighborhoods tend to have high rates of crime and violence that generate low levels of trust and cooperation among residents. Many of these neighborhoods have a history of failed improvement efforts and promises from outsiders that have gone unfulfilled. This context makes it quite difficult to build strong bonds among residents and to build new bridges to the support community.

For Putnam, the most successful local organizations represent indigenous participatory initiatives in relatively cohesive local communities, not those that are implanted from the outside and in neighborhoods with long histories of racial and social division—as was the case with the LISC demonstration sites. Putnam is pessimistic about the possibility of establishing social capital where it does not already exist and where conditions are unfavorable, as is the case in the targeted neighborhoods and many of America's inner cities.[14] For Putnam, the establishment of trust and norms of cooperation requires people to be in contact with each other over a long period of time and to experience firsthand the benefits of social capital. Only with success and continued practice can trust and cooperation be embedded in the local culture.

Temkin and Rohe (1997) question the social capacity building potential of community development efforts based on housing development (such as those attempted by the CDCs organized in the LISC demonstration):

> Developing community is not a high priority for many CDCs. Many CDCs began as advocacy groups that focused on building neighborhood pride and attachment. Over time, however, they have focused on the provision of affordable housing and other tangible outcomes, often to the exclusion of organizing and other activities likely to affect the social capital of an area. (p. 29)

From these vantage points, the LISC demonstration could be seen as a test of Putnam's view of the difficulty of creating bonding and bridging capital, and a test of Temkin and Rohe's skepticism, as well.[15] From Putnam's perspective, it would be very difficult to expect substantial building of social capital using a short-term intervention strategy, such as the LISC demonstration. The LISC demonstration in this light might best be viewed as intended to initiate a long-term process—namely, to begin the long and difficult task of changing existing practices (e.g., conflict, lack of cooperation and trust among residents, distrust of outsiders) that have proven to be detrimental to neighborhood improvement.

Putnam, however, has been criticized for underestimating the ability of newer organizations to foster change and overcome unfavorable conditions (Foley and Edwards 1996). To a degree, the LISC demonstration related some of its program efforts to this potential and built into the program design activities through which newly organized CDCs could experience success, gain momentum, and start to build social capital. Yet, many argue (including critics of the LISC demonstration in the local sites, such as ACORN in Little Rock) that new associations cannot effect change without specifically and directly engaging in political issues and without representing compelling social interests. An important question is whether the CDCs organized in the LISC demonstration (or any CDCs) are able to produce the type of social capital that is necessary to foster significant social, economic, and political change.

Conceptual Model of
Community Development[16]

We now present a structured conceptual model of community development and use it to describe the LISC national demonstration. Much of the model draws on the social capital and network theory outlined earlier. The main purposes of the model are to (1) suggest key elements and relations in the LISC demonstration and in community development more generally and (2) establish a useful framework to analyze the LISC national demonstration. The framework enables consideration of the program design choices made in the LISC program and allows for structured comparison of the LISC demonstration with other community development approaches.

Our conceptual model of community development depicts a set of relations among key elements of program experience. Community development initiatives have many complexities and require multiple decisions within the context of ongoing organizational and program activity; however, their basic operation can be represented fairly simply. Figure 2.2 depicts the main elements of the model. The foundation of the model is that the interaction among program and organization design and implementation attributes, external agents and intermediaries, and local contextual elements yields a set of intermediate outcomes that can eventually lead to more substantial outputs and sustainable community development. The conceptual scheme is not meant to be static or linear;[17] instead, it is intended to help convey the highly dynamic and interactive nature of community development.

Objectives and Intermediate Outcomes

The model posits that the long-term objective of any community development effort is to produce results that are tangible and sustainable. These outcomes can include improvement in the quality of life of local residents (e.g., improved physical infrastructure and affordable housing) and the expansion of employment and other economic opportunities for the targeted population. Although these objectives may be easier to measure than intermediate outcomes, they generally are harder to achieve.

Intermediate outcomes represent both ends and means (i.e., necessary steps for achievement of other intermediate and long-term outcomes). Intermediate outcomes include enhancement of community commitment, capacity, and con-

1. **Program or Organizational Design and Implementation Attributes**
    Geographic and population target
    Mission and goals or strategic focus
    Governance
    Board, staff, and community (resident) influence
    Funding
    Public, nonprofit, or private sources
    Periodic or dedicated
    Matching or direct
    Staff capacity and orientation
    Professional training
    Prior experience
2. **Intermediate Outcomes (enhancement of commitment, capacity, and control)**
    Resident commitment
    Interest in and loyalty to community
    Relationships among residents (i.e., levels of trust and cooperation)
    Vision for community
    Resident capacity
    Leadership
    Financial, technical, and political know-how and wherewithal
    Organizational capacity
    Board development and orientation
    Activities and "spin-offs"
    Implementation of plans
    Staff development and orientation
    Technical know-how (staff)
    Network capacity (i.e., bridging with the support community)
    Public sector, nonprofit, private sector
    Financial, technical, and mentorship relations
    Realization of mutual gains
    Resident and neighborhood control
    Influence of development processes and outcomes
    Power relations with support community
3. **Long-Term Measurable Outcomes**
    Physical and housing development
    Employment and business development opportunities
    Enhancement of resident human capital
    Training and education
    Human service provision

Figure 2.2. Key Dimensions of Community Revitalization

trol and what we call bonding and bridging capital. Our definition of resident commitment includes commonly described measures of social capital, such as

---

4. Local Context
　　Socioeconomic conditions
　　Trust and cooperation among residents
　　Race and class relations
　　City policies
　　Political culture
　　Level and quality of community development activities
　　Competency and capacity of community-based organizations (e.g., CDCs)
　　Nonprofit and private foundation resources and commitment to
　　　community development
　　Private-sector support of community development

5. External Agents
　　Federal agencies and programs
　　State agencies and programs
　　National and regional intermediaries

Figure 2.2. *Continued*

loyalty to community (Temkin and Rohe) and levels of trust and cooperation among residents (Putnam, Keyes et al.). Our capacity definition follows closely from Temkin and Rohe. It indicates the potential for community residents to act on collective commitment, interests, and objectives. It includes (1) individual capacities (e.g., neighborhood leadership, individual's technical and organizational skills); (2) internal neighborhood organizational capacity (e.g., the capacity of newly organized CDCs to undertake real estate development projects); and (3) network or "linkage" capacity (what Keyes et al. highlight and what Putnam and we call bridging capital). These were key characteristics of social capital identified in our review of the literature on social capital and network theory as it relates to community development.

We add resident and neighborhood control to our list of key intermediate outcomes. Increased influence is a desired end-product of most capacity-building efforts. Control is also often the main objective of community organizing efforts, such as the consensus-based efforts in the LISC demonstration. Control indicates the degree to which community residents influence relations, activities, and ultimately outcomes. The integrity of many community development efforts—including LISC's demonstration and comprehensive community initiatives—are dependent on increasing local community control. Finally, and perhaps of most importance, without local control of community development processes and outcomes, it would be very difficult to sustain local commitment and capacity building and to achieve long-term community development objectives.

Local Context and External Agents

In the conceptual model of community development, the efficacy of program and CDC activities is affected by, and in turn influences, the local context. The local context includes neighborhood social (bonding) capital, socioeconomic conditions, race and class relations, network relations (bridges), state and city policies and politics, local political culture, and receptivity of external agents (including potential support organizations) to community development efforts. Some of these contextual factors are more strongly influenced by program activities and relate more closely to intermediate outcomes than others. For example, the pilot site of the LISC national demonstration (the Mon Valley) revealed that the receptivity of the support community can be influenced by program efforts, as can city and state community-development politics. In contrast, socioeconomic conditions and race and class relations are much more difficult to affect directly, particularly in the short term. In many respects, the LISC demonstration was a test of whether the "existing state" of bonding and bridging capital could be substantively altered with an outside intervention.

All three LISC demonstration program sites—Palm Beach County, Florida; Little Rock, Arkansas; and New Orleans, Louisiana—had limited prior experience and success with community development. All had many neighborhoods where social capital bonds and bridges relevant to community development appeared to be lacking. None of the targeted areas had existing CDCs that had demonstrated competency. In all the areas, there was little history of support for community development and there existed few ties between neighborhood groups and the metro-area support community. In all the sites, racial issues affected community development, because most of the neighborhoods eventually targeted for program effort were majority minority (predominantly African American).

This context strongly influenced program design. The program devoted much more attention to educating volunteers and the support community about community development, trying to change attitudes and behavior, and building support for community development, than would have been necessary in more "mature" sites. In addition, priority was given to the recruitment of minority staff. Lastly, expectations and objectives were influenced by the local contexts.

There are also important aspects of the community development context that are not local. In our conceptual model, these are defined as *external agents*. Prominent external agents include federal and state agencies and programs, national foundations, and national intermediaries (e.g., LISC and Enterprise Foundation). In the demonstration program, the primary external agent was LISC.

Program and Organization Design

Program and organizational design attributes reflect distinct objectives and strategic choices that can influence intermediate outcomes and the achievement of long-term goals (i.e., measurable outcomes and sustained community development). The geographic and population target can effect the basic mission and eventually the success of community development efforts. The range of target options includes residential neighborhoods, business districts, distinct economic and racial groups, and combinations of these. Relevant criteria related to targeting include need (serving those with the greatest need), triage (directing assistance to those who can benefit the most from the help), cohesiveness (working with a cohesive group or area), diversity (e.g., geographic area with mixed land use), and political expediency.

In the LISC demonstration, there were two levels of targeting: site and neighborhood. At the site level, metropolitan areas were selected that had a prior lack of concerted community development effort, yet, as indicated by site assessments, demonstrated need, commitment, and potential to change. The second level of targeting was at the neighborhood level in selected sites. The targets were low-income residential neighborhoods that had both clear need and strong potential for community engagement (e.g., potential for leadership development) and change as suggested by interviews undertaken by community organizers.

The mission and goals of community development interventions are affected by their target area or population and, in turn, influence the selection of the target. The mission of community development can be very narrow and specific (e.g., rehabilitate rental housing, increase local businesses, expand employment opportunities for residents); emphasize different elements of community development (e.g., housing, economic development, advocacy or political change); be more tangible (e.g., housing development) or less tangible (e.g., increasing social capital) or include elements of both (like the LISC demonstration attempted); or be much broader, even holistic, in nature, capturing a range of elements as do the comprehensive community initiatives (see Chapter 3). Missions can remain fixed, evolve over time, or shift with changing circumstances and contexts.

In the LISC national demonstration, the initial objective as we describe it (and perceive it) was to enhance community development commitment, capacity, and control where it was previously lacking. More specifically, the main objectives were to (1) establish community-based and -controlled CDCs with neighborhood leaders as board members; (2) foster beneficial financial, technical, and political contacts between residents of targeted neighborhoods and leading organizations and individuals in the support community; and (3) have CDCs (and CDC volunteer board members) develop and then demonstrate their competency through the completion of housing development projects. All of these objectives can best be characterized (in the conceptual model) as intermediate outcomes.

The character of demonstration program objectives reflects the nature of the local contexts in which the LISC demonstration was undertaken. The achievement of "harder" measurable results was not realistic during the demonstration period (two to three years), but progress on the intermediate outcomes could provide the foundation in each site for sustained effort and eventually for more extensive tangible results. Yet, even progress on the less tangible intermediate outcomes would be difficult, as suggested by the discussion of social capital, particularly the pessimistic view of Putnam with regard to the possibility of creating social capital where it has not existed. Housing development per se in the demonstration was a narrow end and could be criticized as such (Temkin and Rohe). However, it was also a means to larger ends, that is, enhancing local commitment, capacity, and control of community development.

The governance structure of a community development effort can take on a variety of forms. A key dimension is the degree of effective target population (e.g., resident) influence. Influence is primarily affected by the position and prominence of the target population and others on committees and boards, the nature of target-population interaction with program staff, and target-population engagement in planning and in setting missions and agendas. Community development activities can be controlled by their target populations, staff members (e.g., executive director), boards, funders, or some combination of these. Temkin and Rohe argue that one of the key elements of social capital is the ability of residents to act. That is why we focus on changes in community commitment, capacity, and control in our analysis.

In the LISC demonstration, the two most influential institutional agents were LISC and the CDCs created by the initial organizing efforts. In each site, as is their standard practice, LISC established a Local Advisory Committee consisting of major program contributors. The LISC Advisory Committee's main responsibility is to oversee the work of the local LISC program director and to review grant and loan requests from CDCs. The local program director is hired (and fired) and supervised out of LISC's national office in New York City.

One of the main objectives of the demonstration was to ensure that the newly organized CDCs were rooted in, and responsive to, the community. Organizing efforts were meant to guarantee that CDC boards fostered new neighborhood leadership, represented the diverse interests of targeted residential neighborhoods, and were truly controlled by volunteer board members.

The program design envisioned two basic options at the end of the demonstration period (i.e., at transition). One option (and the one preferred by Eichler, based on his experience in the Mon Valley) to achieve resident control of community development was the establishment of coalition organizations, composed of and controlled by the CDCs. The coalitions were intended to increase resident control of community development through the institutionalization of new ties and relations across (LISC-organized) neighborhoods that shared common problems and interests and could work together on advocacy efforts, as well as share expertise and staff. The other option was that the CDCs would operate independently, presumably with their own offices and staffs, like CDCs in other LISC sites. The official program description was that at the end of the demonstration period, the CDCs and their supporters in each site would decide what would work best; that is, there was no fixed transition design.

Funding is arguably one of the most critical attributes of community development efforts. By its very nature, community development (as described here) requires the infusion of funds (to geographic areas or population groups lacking resources) from outside the target area. The key dimensions of funding include scale, uses, sources, flow, and contingencies. Funding can be of large scale or modest proportions, for general purposes or specific ones (e.g., for operating expenses or project activities), and from single or multiple sources. The funding source(s) may be public, private, or nonprofit. The stream of funding can be "one-time," annual, or longer. Funding can be without any specific requirements, or it can be contingent on matches, performance, or both.

The main source of local funding in the LISC demonstration was the private sector, including foundations, financial institutions, and corporations. Some of the money (approximately one-fifth) was dedicated to the initial organizing efforts (mainly staff salaries). The remainder of funds capitalized the LISC loan and grant pool, with most of this money targeted for use on housing-project loans. As is standard LISC practice, CDCs were required to make individual requests to secure funds and then needed to report back to local LISC program staff on the use of the funds.

Staff capacity and orientation (i.e., professional training and experience) can also have a strong influence over program and organizational direction. The size and the professional training and experience of staff will affect the types of activities that CDCs engage in and the scale at which they operate. These attributes will also affect the relative influence of staff, the target population, and board members.

In each site, the program established a development team that included a local coordinator and three community organizers; their tasks were to organize the new CDCs and help develop the capacities of the resident volunteers. The key criteria used to hire local coordinators were that they be smart, hold a graduate degree in a field related to their work (e.g., social work, public policy), be sensitive to a broad range of community development issues, show commitment to the LISC/consensus approach, and be willing to learn and develop on the job under the guidance of Michael Eichler. Prior experience with community organizing and real estate development were not requisite. Only in New Orleans was strong priority given to hiring someone local as the coordinator.[18]

The desired characteristics for organizers were similar to those for local coordinators, but without the same emphasis on formal education and with greater emphasis on (1) connections to the neighborhoods they would be working in (e.g., similar ethnic or racial background as residents) and (2) facility to learn on the job (under the direction of local coordinator).

The final staff position filled was the local LISC program director. The program directors at the demonstration sites were selected primarily using standard LISC criteria. This included that directors have experience in housing finance, interact well with the private sector, and have the ability to work well with community residents.

Notes

1. Wilson (1987) first made "social isolation" from job networks and "mainstream role models" a major issue in the urban poverty equation. However, this motivated relatively little interest or research directly addressing social capital, and the spatial organization of social capital, until Putnam's (1993) and Wilson's (1996) subsequent work (Briggs, 1998, p. 2).

2. Putnam's work summarized here includes *Making Democracy Work* (1993), "Bowling Alone" (1995a), "The Strange Disappearance of Work" (1996), and *Tuning In, Tuning Out: The Strange Disappearance of Civic America* (1995b).

3. The term *social capital* was coined in 1961 by Jane Jacobs. It was further developed most prominently by James Coleman (1988). Putnam's social capital construction is a slight deviation from Coleman's definition of social capital in *Foundations of Social Theory* (1990).

4. For Putnam, "indicators" of social capital include (as he and his team of researchers measure using the national General Social Survey) voter turnout, newspaper readership, participation in voluntary associations, attendance at community meetings, religious membership, political clubs, labor unions, literary or art discussion or study groups, and sports clubs and organizations.

5. Putnam's views have been subjected to considerable criticism including the following: that there is counterevidence to the decline of civic engagement between 1965 and 1995 in the United States—namely, that this has been an era of unprecedented advances in political and public engagement, for example, in women's rights, gay and lesbian liberation, consumer and environmental movements, pro-life and pro-choice movements, and community development corporations; that Putnam fails to account for the "evolution" from local to national and transnational associations; that some forms of association have changed and Putnam does not pick this up (e.g., membership shifting from YMCAs to private health clubs); and finally, that he does not recognize unique aspects of social capital in low-income and minority neighborhoods.

6. In the LISC demonstration, there was much less attention to building new bridges with governmental and public agents. Public policy and governmental agencies in the national demonstration were mainly used for general support of efforts and as a financial resource for community development activities pursued by the CDCs. Putnam would suggest (and we confirm in Chapter 8)

this as a program weakness, because social capital is not a substitute for effective public policy, but rather a prerequisite for it and in part a consequence of it.

7. In fact, there are instances where too much bonding, but too few bridges, has proven destructive. For example, the social capital found in some white ethnic neighborhoods of America's cities, such as the Bensonhurst section of Brooklyn or Howard Beach in Queens, has helped make such places rather unfriendly, and at times deadly, for some outsiders, especially minorities (Hornburg and Lang 1998). The Michigan Militia have strong group cohesiveness (and probably even bowl together); however, their social capital has not been used to serve broad interests.

8. Temkin and Rohe (1998) analyze neighborhood change in Pittsburgh between 1980 and 1990 using measures consistent with the two constitutive elements of social capital, as well as other (control) variables that have been used to explain neighborhood change in previous research. The social capital constructs are defined with principle component analysis of detailed neighborhood (census tract) data. The effect of social capital on a neighborhood's trajectory is estimated econometrically, with changes in property values as the dependent variable and social capital constructs and control variables as explanatory (right-hand side) variables.

9. Temkin and Rohe's (1998) sociocultural milieu is an expansion of Warren and Warren's (1977) identification of three critical social-structural characteristics in community development: identity, interaction, and linkages.

10. Leveraging requires forming effective alliances with actors outside the community whose decisions will affect the neighborhood's character over time.

11. Nonredundant contacts are particularly valuable because they help parties gain access to new resources and opportunities. In contrast, redundant ties are duplicative—that is, with the loss of these ties there would be no loss of resources or opportunities, as other ties would provide access to these. Redundant ties are viewed as inefficient. A criticism of Burt is that he did not fully recognize (or appreciate) the value redundant ties may play in helping to build social capital of the bonding type.

12. It has been argued that some local public and nonprofit community development agencies and professional staff act in a mainly self-interested manner as entrepreneurs filling structural holes between the residents of low-income communities and outside resources, mostly for their own benefit and marginally for the benefit of the intended clients (R. Gittell and Thompson 1996).

13. The "filling-structural-holes role" of CDCs is consistent with the way many well-established CDCs view themselves. Interviews of executive directors of some of the largest CDCs in the country reveal how they view their CDCs as "gap fillers," finding weak spots in the community and shoring these up (LISC 1994).

14. Putnam found that state-level differences in social trust and group membership are substantial and closely interrelated and reasonably stable at least over the period from the 1970s to the 1990s.

15. The discussion of Putnam suggests both the limits and possibility of solving some of our nation's urban problems through relatively low-cost association-strengthening local initiatives. In sum—using Putnam's vernacular—the LISC demonstration tried to increase social capital within and across low-income communities, as well as between low-income communities and the metropolitan-area support community. The social capital to be created through the LISC demonstration was mainly of the bridging sort—bringing together people, and people with resources, that normally do not come together. These bridges were intended to serve both individual community interests and broader metropolitan-area interests.

16. This conceptual model of community development was first developed by the authors with Margaret Wilder on work undertaken for the Lilly Endowment in the summer of 1995. The framing of community development has evolved inductively. The design of the conceptual model benefited

from the collective research and professional experience of the three contributors, including the preliminary work and interim documentation related to this report and an assessment of a Lilly Endowment community development initiative in Indiana's Enterprise Zones (R. Gittell, Vidal, and Wilder 1997).

17. For example, funding may be viewed (in the same program context) as part of program design, an intermediate outcome and a measurable long-term outcome.

18. In New Orleans, the local culture and community development context is such that priority was given to hiring a person of color with local experience. New Orleanians tend to put greater trust and faith in "one of their own" than the other two sites, which are more open to outsiders. In addition, the pace and local style are quite distinct from those in other cities.

CHAPTER 3

A Social Capital
Perspective on Community
Development Practice

This chapter draws on the theories and framework introduced in Chapter 2 to examine different community development practices. We highlight four major activities in the field: the CDC movement, comprehensive community initiatives, the federal Empowerment Zone and Enterprise Community program, and community organizing. The discussion is intended to both position the LISC demonstration relative to other efforts in the field and use a description of a broad range of practices to provide informed expectations about what the LISC demonstration could reasonably be expected to accomplish.

Community Development
Corporations (CDCs)

The CDC movement is the most relevant to our understanding of the context for and dynamics of the LISC national demonstration program (thus, we devote more attention to this activity than the others). That movement had its origins in the War on Poverty and the civil rights movement in the 1960s, and grew in the 1970s as low-income neighborhoods organized themselves either to oppose threats to their physical integrity (e.g., partial demolition to make way for construction of urban links in the interstate highway system) or to take advantage of new federal programs. Starting in the early 1980s, however, the rate of growth in the number of CDCs increased dramatically. The large national financial intermediaries (e.g., LISC and the Enterprise Foundation) have played a central role in that growth story, and the demonstration program is part of it.[1]

Housing as a Point of Entry
Into Community Development

The production of assisted housing is the activity for which CDCs have become best known and for which they have attracted the greatest financial and political support. Ninety percent of all CDCs are engaged in housing production (NCCED 1995). Walker (1993) estimated that CDCs produce about 23,000 units of housing a year—roughly 13 percent of all federally subsidized units built annually (a combination of new construction and rehabilitation of existing dwellings).

Most CDCs formed in response to an immediate perceived need or problem in the community, and hence engaged in one or more types of *nondevelopment* (i.e., not physical development) activity—for example, organizing, advocacy, neighborhood cleanup, or starting an after-school program for youth—before taking on their first (physical) development project.[2] Housing has been the area in which most groups get their start in development.

Using housing as a point of entry has helped the CDC movement grow (Vidal 1997). Up until the early 1980s, when Congress stopped funding the federal Section 8 New Construction and Substantial Rehabilitation Program, housing was significantly less difficult and less risky than other types of development CDCs might wish to undertake, such as small-scale commercial or industrial real estate development or business development (Vidal 1992; Vidal, Howitt, and Foster 1986). The loss of project-based Section 8 subsidies, and especially the switch in 1986 to the low-income housing tax credit as the major form of federal support for affordable housing, made housing development both more complex and more risky than it had been (Bratt et al. 1994). Even so, it remains the case that starting with housing reduces a CDC's difficulty in raising needed financing and increases the likelihood that its early activities will be successful, thereby establishing its reputation and positioning it well to move on to larger projects or to other types of activity.

The scale of CDC housing production varies greatly across organizations, partly because local housing conditions vary, and partly because groups entering the field typically start with small projects and then move on to larger, more ambitious ones. The great majority of CDCs produce at a very modest rate: one-half of all CDCs active in housing produce (or sponsor) fewer than 10 housing units a year. In contrast, 10 percent of CDCs—typically the largest and most prominent ones—produce more than 50 units per year; this is roughly twice the number produced annually by the small, independent, for-profit developers who constitute 75 percent of U.S. homebuilders (Vidal 1992; Walker 1993).

Because the LISC national demonstration program was designed to spawn new CDCs, the experience of new CDCs (rather than CDCs in general) is useful in shaping reasonable expectations about what the program should be expected to accomplish. As the movement has grown and matured into a fledgling industry, more and better support has become available to new organizations. CDCs founded in the 1980s and 1990s are more likely than their predecessors to make real estate development their initial activity, and the average time these groups take to get their first projects under way has declined from almost six years to about nine months (Vidal 1992). Once a development project has been identified, LISC's experience, based on its work in cities where at least the beginnings of a support system are in place, suggests that moving it to completion will take between one and two years. The LISC demonstration sought to organize new CDCs that would move immediately into development, would produce products without delay, and would do so in locales that offered minimal support beyond that offered directly by the program. Viewed from this perspective, the program was quite ambitious.

Although housing development is their most widely shared activity, most CDCs engaged in other community improvement activities, too; the mix chosen by any particular organization depends on perceived community needs, staff capacity, and the availability of funding and any needed technical assistance. For example, more than 75 percent of CDCs provide resident services such as homeowner and tenant counseling, weatherization assistance, or housing for homeless people. More than 70 percent conduct advocacy and organizing activities and more than 60 percent provide some type of human services (e.g., emergency food pantry, job placement, child care, teen pregnancy counseling). Twenty-five percent are active in small-scale commercial and industrial real estate development, and a similar number do small business lending. These programs address pressing community issues, and many CDCs have had a broad notion of community improvement at the core of their mission, even if housing was an early area of focus.

CDCs as "Gap Fillers"

Support for the growth of CDCs derives not only from a growing acknowledgment of their accomplishments (NCCED 1995) but also from the gradual withdrawal of other entities working on behalf of poor communities. This is very clear in the area of housing. With the demise of deep-subsidy federal production programs, developing affordable housing in poor inner-city communities has become financially unattractive to most for-profit developers. In many places,

CDCs are the only organizations willing and able to assemble the many sources of funding—so-called "creative financing"—necessary to produce low-income housing (Vidal 1995).[3]

More broadly, their supporters frequently cite the fact that CDCs are doing the difficult job of providing service and leadership in communities that need help and that other agencies cannot or will not serve: They have the ability to operate in complex environments where others can't, and a willingness to do projects that others won't because the projects are too small or too risky (Vidal 1992). It is not surprising, then, that many CDCs think of themselves as "gap fillers"—responding to community problems that no one else is willing to take on (Leiterman and Stillman 1993). Alternatively stated, CDCs see themselves not only as vital institutional infrastructure for their communities, but as agencies moving in to fill structural holes (in the sense described by Burt [1992]) created by the disappearance of other community actors.

■ Reliance on External Resources and Partners

Because CDCs work in poor, disinvested neighborhoods, they make much of their impact by leveraging resources from sources outside the community. Their supporters include government agencies, intermediaries, foundations, banks, education and training institutions, trade associations, and technical assistance providers. When the support system is at its best, the three key types of support—funding, technical assistance, and political support—are designed and coordinated into programs that meet the particular needs of local CDCs and their communities.

The major sources of financial support for project development are federal, state, and local government; foundations; and banks and corporations. The public sector typically makes its funding available directly to the CDC developer.[4] Other partners commonly channel their dollars through national intermediaries or local partnerships. Technical support is most important to smaller, younger groups that have not become large enough to hire specialized expertise in-house or to help groups of any size expand into an unfamiliar line of activity.

Many members of the support system are in a position to provide less tangible—but quite valuable—kinds of assistance. These include networking, advising, coordinating activities and resources to make them readily available and mutually supportive, building a local culture conducive to charitable giving, and advocating community-based approaches as the preferred vehicle of social and economic development. Local governments in particular are in a position to provide support in a variety of forms, including some not available from other sources—such as tax abatements, low-cost land and buildings owned by the city (e.g., because prior owners failed to pay property taxes), and project-related infrastructure.

Cities and states—but particularly cities—have been expanding their support for CDCs, providing them with a mix of project financing, administrative funding, predevelopment loans, and technical assistance (Goetz 1993:122). These types of support are commonly tied to locally designed programs. In places like Boston, Chicago, Cleveland, Minneapolis, New York, and San Francisco, where CDCs have become strong and numerous, supportive institutions have learned to work effectively both with the CDCs and with one another, providing complementary types of assistance and coordinating their activities.[5] In other places, particularly cities in the South and Southwest (e.g., Miami, Dallas, and New Orleans), this learning process is more recent, and support for CDCs is more fragmented and uneven—a central issue that the LISC national demonstration program sought to address.

■ The Role of the Intermediaries

Partly in response to federal withdrawal and partly in response to growing recognition of the efficacy of CDCs, CDCs' institutional supporters—and the financial and political support they bring to bear—have increased rapidly in number and sophistication during the 1980s and 1990s. The national financial intermediaries have played a central role in this expansion. Functioning as specialized community development "banks," the intermediaries receive grants and low-interest loans from foundations, banks, corporations, and public-sector agencies, and use the resulting financial pool to provide grants, loans, and credit enhancements to CDCs and selected other nonprofit housing providers.

The success of the intermediaries derives from their ability to meet the distinctive institutional needs of both financial supporters and community organizations. CDCs work in weak markets where property values and resident incomes can support limited debt; this makes CDCs' projects riskier investments than most conventional development projects. Intermediaries facilitate risk sharing among financial participants, enable them (especially those from the private sector) to reduce their exposure to risk, and help them take advantage of federal tax credits. Funders benefit from the intermediaries' experience and expertise in assessing risk and in structuring projects and financial packages that avoid unnecessary exposure and benefit from public-sector support. Funders also save the administrative costs of dealing with individual CDCs and evaluating individual funding proposals (including the substantial costs that would be entailed in learning how to do so). With their successful track record, the intermediaries serve as quality control guarantors, providing a "due diligence" for the projects in which they participate (Vidal 1996).

For their part, CDCs benefit from the technical and program design expertise of the intermediaries, from their ability to transfer lessons learned in other localities, and from their ability to attract funding to the field. For example, national and local intermediaries have helped CDCs to tap substantially increased levels of support: Total grant support to CDCs from foundations and corporations was $179 million in 1991 (the last year for which figures are available)—up 72 percent from only two years earlier (Council for Community-Based Development 1993). The success of LISC and the Enterprise Foundation in mastering the complexities of working with the low income housing tax credit has allowed them to raise more than $7 billion in equity capital for housing developments (Vidal 1997).[6] In addition, the national intermediaries and the organizational networks they represent have undoubtedly been an important force behind the federal government's moves to give CDCs and other nonprofits a growing role in providing affordable housing.

Importantly, the major national intermediaries have often acted as catalysts in forming local intermediaries, sometimes in the form of local housing partnerships that broaden the base of local financial and political support for community-based development and take on a locally developed support agenda. One of the oldest and best known is the Metropolitan Boston Housing Partnership, whose members include major banks, insurance companies, and utilities, and other local businesses; the city of Boston; the state housing finance agency; local universities; and local housing and community development organizations. Although created to support low-income housing production, it has become increasingly concerned about and involved with helping participating CDCs to address housing-management problems and related tenant-services issues (Vidal 1992:132-133).

CDCs as Voices for Their Communities

CDCs view themselves as representatives of their communities, and external agencies such as local government, banks, and the intermediaries commonly accept them as speaking on the community's behalf. Their supporters commonly credit them with articulating a vision of the community's future, helping local people help themselves, and using their activities to strengthen the sense of community itself (Vidal 1992). Most of those that are active in housing development are recognized as Community Housing Development Organizations (CHDOs), which means that they are eligible for HUD funding because they meet federal criteria regarding community representation on their boards of directors.

Nevertheless, some critics concerned with governance issues and community empowerment have raised questions about whether CDCs are adequately representative of, and accountable to, their communities (see, e.g., M. Gittell, Newman, Bockmeyer, and Lindsay 1996). Some of these concerns are a response to the fact that professional staff members often have considerable influence on CDCs' agendas. Others focus more on the composition of CDCs' boards, which often include some members who neither live nor work in the neighborhood.[7] Concerns of this kind were important in shaping the LISC national demonstration, including the choice to have the new CDC boards be composed solely of neighborhood stakeholders and Michael Eichler's interest in encouraging the CDCs to form a coalition through which they could share professional staff— thereby keeping the task of setting each CDC's agenda firmly in the hands of its resident board.

Lessons for the Assessment of the LISC National Demonstration Program

The broad outlines of the demonstration program were a logical extension of current community development practice and LISC's prior experience. The approach of having new CDCs and new local systems of support for their work start with small housing developments was well established in the field, and LISC had extensive experience with it. This gave the program great credibility with both funders and community residents. At the same time, even residents whose top-priority neighborhood issue was not housing could readily appreciate its usefulness as a way to establish a reputation that would help them move on to address other pressing issues. Given the field's recent experience, the initial two-year program timetable—moving from selection of an initial group of resident volunteers to development projects well on their way to completion— was very optimistic; meeting this goal essentially required that everything go according to plan. The revised three-year timetable was demanding but more realistic.

More broadly, LISC framed the program's goal in terms of bringing each site into the mainstream of the community development industry. The near-term program goal was to establish six local CDCs in each site and help each group to successfully complete an initial development project. In the long term, however, the goal was to generate the sort of community development capacity, including a solid system of local support, that LISC had helped to foster in other localities: bridging and leveraging forms of social capital with the four dimensions described by Keyes et al. (1996). Just as the new CDCs can be thought of as agencies filling Burt's (1992) structural holes in their neighborhood, LISC can be seen as seeking to help fill such a hole in each of the demonstration sites.

Comprehensive Community Initiatives and Community Development [8]

Comprehensive community-based change initiatives (CCIs) have mushroomed in number since the early 1990s. By some counts there are now more than 40 such initiatives around the country. Although they vary greatly in approach and structure, they share a joint focus on people *and* place. They typically seek neighborhood transformation, and sometimes also the reform of city-wide or metropolitan-wide systems (e.g., the job training system or the local school system). Hence, they aspire to address multiple aspects of neighborhood life, including housing, economic and physical development, services (e.g., schools, human services, youth-serving activities), and safety (Brown 1996; Kubisch et al. 1997). In addition, all CCIs attach importance to "community building"— strengthening bonds among community members and increasing community capacity. This holistic approach is a recognition that the problems facing poor neighborhoods are many and that the many dimensions of family and community life are interdependent. It also reflects the hope that addressing several dimensions of community life in a coordinated way will be synergistic.

The revival of interest in comprehensive approaches is a response to the convergence of two streams of programmatic experience.[9] On the one hand, it represents the conviction of philanthropic funders that enough is now known about how to improve the lives of disadvantaged families to make real improvement possible. According to this view, improvement can occur if enough resources can be brought to bear through a process that responds to perceived community needs, values, and aspirations—the hard-to-measure qualitative elements that one expects to accompany Temkin and Rohe's (1998) measured elements of social milieu. On the other hand, it is a product of twin disappointments: first, that the "bricks and mortar" emphasis of many CDCs, by itself, too seldom translates into sustained community improvement in social and economic terms; and, second, that human services—typically categorical and focused on individuals or, at best, families—produce improvements that are modest and easily overwhelmed by the conditions of life in poor neighborhoods (Kubisch et al. 1995). In Briggs's (1998) terms, this latter concern can be understood as a frustration that categorical programs targeted at individuals or nuclear families do not help their beneficiaries strengthen the system of social supports (the coping type of social capital) that ensures their ability to at least "get by." As a result, programmatic gains in linking assisted families or individuals to opportunities (whether inside or outside the neighborhood) that would help them "get ahead" are eroded.

Two Examples

Although CCIs generally have these core principles and attributes in common, they nevertheless exhibit great variety. Two sharply contrasting examples illustrate both the range of initiatives that fall under the umbrella term "CCI" and some of the features these groups have in common.

■ The Neighborhood and Family Initiative

The Neighborhood and Family Initiative (NFI) began in May 1990 under the auspices of the Ford Foundation. Ford funds the community foundation in each of the four sites—Milwaukee, Hartford, Memphis, and Detroit. The community foundation, in turn, takes the lead in assembling a group to work on the problems facing a single target neighborhood and is the agent through which NFI funds flow to local projects and programs. Ford also funded the Center for Community Change to provide all the sites with technical assistance.

Sites were given broad latitude to structure their local program, identify areas of focus, and set priorities within the context of a few guiding principles. Substantively, they were charged with the task of engaging in a strategic planning process that would make a comprehensive assessment of neighborhood conditions and seek to develop projects and programs that would take into account the interrelations among the physical, economic, and social aspects of family and community life. Its goal is "to test whether investments are maximized when they occur together in single neighborhoods and simultaneously target the whole family, the physical and economic environment, and the community's leadership and organizational needs simultaneously" (Chaskin and Joseph 1995:100).

This very early CCI is distinguished from previous *coordinated services* efforts in three ways, two of which have social capital building aspects. First, it emphasized the importance of enhancing neighborhood leadership and mobilizing broad participation—clearly an effort to strengthen connections among community members and, through leadership development, improve the quality of the neighborhood's institutional infrastructure. Second, it sought to "bridge the traditional separation between human services and physical revitalization, including housing and economic development" (Chaskin and Joseph 1995:100). While this goal is stated in programmatic terms, any successful effort to pursue this objective necessarily builds bridging capital both within the neighborhood (between organizations engaged in different types of activities) and in the support community (between the funders and program designers that support those different types of activities). Third, it sought to move beyond the simultaneous operation of parallel programs within a neighborhood to "more organic connections among programs that build upon one another and add up to more than the sum of the parts" (p. 100).

A more explicit mechanism for creating bridging capital is the requirement that the initiative be governed by a collaborative that includes both leaders from the target community and a variety of other well-regarded citizens from the public and private sectors who can contribute to the neighborhood revitalization process by virtue of their positions, experience, or perspectives. Residents are seen not simply as beneficiaries of NFI, but as key collaborators in its design and implementation. The underlying assumption of the initiative is that by

array of contrasts to the approach taken by CDCs and their supporters in general, and by the LISC national demonstration program in particular.

■ Comprehensive and Holistic Intent

As noted earlier, CCIs are comprehensive and holistic in intent. However, Kubisch et al. (1997) argued that CCIs are best understood as being comprehensive in the sense that they are *enabling*; the long-run goal is to address many facets of community life, but local participants have broad latitude to set priorities, devise strategies, and choose how they wish to begin based on local problems and opportunities. This program design contrasts sharply with the LISC demonstration program; given its short time frame, LISC set comparatively modest, focused goals and asked participating communities to agree to make their first undertaking a real estate project (preferably a housing development). At the same time, the LISC demonstration presented that modest goal as a "ticket to the table" that would enable the new CDCs to set their own agendas, and this aspect of the program was critical to maintaining volunteer motivation and commitment.

Actual CCI experience in the early years often differs less from the LISC demonstration than the differences in goal statements might suggest. Even CCIs with a very broad vision must start actual program activity with a limited number of specific projects or programs, especially if the program staff is small (as it usually is—often a single individual). As a result, CCIs tend to focus on a small number of functional areas in their early years, and projects that have a tangible presence and visible impact in the community have considerable appeal, particularly among community residents. These early efforts are sometimes the first steps in implementing a strategic plan, but they may also be a response to strongly felt local needs or to the development of a particular opportunity, which means they may move forward with little, if any, coordination (Kubisch et al. 1997). Nevertheless, the comprehensive approach does allow local participants to pick where they want to begin, rather than following a prescribed program.

Even when CCIs mature, however, there are to date few signs of the hoped-for synergy among the programs they sponsor. The idea of comprehensive, integrated development has provided a lens through which each NFI collaborative has considered its work; "in general, however, program development has followed parallel categorical streams of activity" (Chaskin, Dansokho, and Joseph 1997:5). In fact, within NFI, only two of the four NFI sites actually developed a single, over-arching strategy. Detroit funds activities in three programmatic areas (child and family services, employment and economic development, and housing and physical revitalization), but has sought to do so in a way that strengthens the linkages (bonds) among individuals, families, and the numerous organizations active in the target neighborhood. The Milwaukee NFI is focused on providing neighborhood residents with "livable wages" through employment and economic development programming. Equally important, program structure has been strongly influenced by the availability of local opportunities, networks of association that provide access to those opportunities through collaborative members, and the need to act within particular funding periods (Chaskin et al. 1997)—all of which mitigate against a focused, coordinated, strategic approach.

■ Significant Funding Concentrated in
  a Small Number of Neighborhoods

CCIs concentrate substantial funding on a small number of target communities—sometimes only one, and rarely more than half a dozen; NFI and CCRP are both typical in this respect. Although the resources available to the initiatives arc inevitably small relative to a comprehensive conception of community need, they are significant. For example, NFI has provided each of its four collaboratives with $9 million over the program's seven years, and CCRP raised over $9 million to support work in five neighborhoods. In contrast, total funding for the LISC demonstration program was approximately $5 million, to support a program active in 18 neighborhoods.

The funding available under the large CCIs is substantial enough to attract the attention and participation of a range of local actors, and the clear hope is that direct philanthropic funding will ultimately leverage participation by others if the program develops a vision and approach that make sense and are the outcome of a broadly collaborative process. The funding is quite flexible, and sponsoring foundations seek to assure participating localities that as long as they remain committed to the initiative and show substantive progress, they can count on support for a sustained period (e.g., five to ten years). All parties acknowledge that the severity and tenacity of the problems being addressed mean they must be prepared to "stay the course" until important results are apparent and the program develops a stable base of local support.

Sustained CCI effort over a period of this length clearly requires, at a minimum, a high level of social capital as Putnam (1993) originally conceived it: trust and cooperation among individuals engaged in long-term relationships. However, we know little about whether and how CCIs successfully build such capital among prospective supporters of community development. Evaluators of CCIs have not applied a social capital framework and have focused much of their examination of participation patterns on the role of residents, which varies widely,[10] and of "bridging" individuals (members of the target community who have respected roles outside the community, e.g., as public servants or leaders of well-recognized nonprofit organizations)—one embodiment of bridging capital (Chaskin and Joseph 1995).

commitments of additional dollars from state and local government and the private sector.

Despite these surface similarities, the zones' strategic plans and activities exhibit great variety, because localities were encouraged to exercise broad latitude in identifying local problems and opportunities, and in crafting strategic responses appropriate to the local context. For example, all zones seek to promote economic opportunity for zone residents, but the relative emphasis they place on creating jobs (through entrepreneurship and business development) versus preparing residents for jobs varies considerably. Even within these well-defined categories, different zones take very different approaches. Some take a focused occupational-training tack; others are working toward a full spectrum of services ranging from literacy or basic education and job readiness to postplacement follow-up support, including help with day care and transportation. More broadly, some zones have targeted the bulk of their efforts on increasing economic opportunity, whereas others are pursuing economic opportunity goals within a very broad framework that emphasizes the community's quality of life (Wright et al. 1996).

Community-Based Partnerships and a Strategic Vision for Change

Within the EZ/EC program, issues directly relevant to the LISC demonstration program and to enhancing social capital come into play, primarily in the context of two of the key principles: community-based partnerships and strategic vision for change.[14] The strategic planning process provided the point of departure for both. Zones that took this process seriously conducted broad outreach to major neighborhood constituencies and helped residents and other stakeholders play a substantive role in shaping the strategic plan. In so doing, they presumably strengthened existing weak ties (some of which are included in Temkin and Rohe's [1998] indicators of social milieu) and increased their number, thereby making internal neighborhood networks denser and enhancing the community's bonding capital. Zones that did a good job of leveraging financial commitments from public and private entities outside the zone began the process of cultivating new working relationships (building bridges) that would better link the zones to their metropolitan economic and political systems. In both cases, a good strategic planning process lay the foundation on which the program itself could build.

■ Community-Based Partnerships

Community-based partnerships can be seen as vehicles for generating social capital in three distinct ways. First, they can create and strengthen organizations and collaborations within the community, enabling community members to participate in aspects of community development and community life that concern them by setting priorities, developing programs, and participating in them. In this respect, they provide avenues of opportunity for CDCs and other community-based organizations to develop new contacts, access new resources and opportunities, and gain greater influence over what happens in the community. In network theory terms, they not only strengthen weak ties but also enhance the institutional infrastructure of the community, thereby enhancing its ability to take effective action on its own behalf.

Second, the EZ/EC program guidelines emphasize the importance of linking zones and their residents to external sources of resources and opportunities. This includes helping residents find employment outside the zone, but it also encourages zones to encourage better access to, and participation in, government and nonprofit agencies by zone residents, giving them more say in decision making that affects their lives and communities. Partnerships in which the zone community has a meaningful voice are seen as the principal mechanism for accomplishing this. Many zones seek to promote this goal via the zone's governance structure, including representatives of key state or local government entities and of the private sector (as well as community members) on the zone's governing board. Another common bridge-building strategy is to involve private companies (typically either large employers or representatives of industries with particular labor needs) directly in the design and implementation of job-training programs. No matter what partnership vehicles are chosen, however, the goal is clearly to create bridging or leveraging forms of social capital.

Finally, designers of the federal program clearly intended that the partnerships created would increase the degree of coordination and collaboration among public, private, and nonprofit entities in support of zone-improving projects and programs. Participants are encouraged to effect greater cooperation among government agencies at all levels, resulting in faster, more efficient use of public resources, and increased private-sector participation, including more private support for community initiatives.[15] This aspect of the program clearly seeks not only to stimulate the use of EZ/EC funds to leverage capital from other sources, but also to encourage local jurisdictions to build a support structure for the revitalization of the zone. The four elements of social capital needed to support community development identified by Keyes et al. (1996) provide one way to think about this task.[16]

■ Strategic Vision for Change

The principle that each zone should have a strategic vision for change reinforces the theme in the strategic planning process that the community needs a shared vision of its future to guide its efforts during program implementation. Less obvious is the fact that sustaining a strategic plan of action over time requires that one person, or a small group, assume responsibility for keeping the work focused on pursuit of the strategy, for adjusting tactics as conditions and opportunities change, and for adapting the strategy itself in response to observations of what works and what doesn't. For this reason, EZs and ECs typically identify a lead agency through which public-sector program funds are channeled and distributed.[17]

In some cities, this body is a de facto agency of city government—often incorporated as a nonprofit organization so it is eligible for foundation grants and able to operate free of the procedural guidelines (e.g., civil-service position descriptions and work rules, contracting procedures) that bind most city governments. In other places, newly created independent nonprofit organizations are taking the lead implementation role under the guidance of a board of directors that represents a variety of stakeholders. This latter path holds greater promise for giving zone residents a meaningful voice in shaping zone programs, and hence for increasing resident capacity and strengthening the zone's institutional infrastructure.

Nascent Lessons

Like CCIs, EZs and ECs have been slow getting started. Localities began preparing their applications for the zone program very early in 1995, but unlike participants in CCIs, they had only a few months to conduct their strategic planning process, even though that process was supposed to involve substantially more local residents than most CCIs include. Partly for this reason, many zones (especially EZs) have had to do considerable additional planning before actually getting their programs under way.

Unlike privately funded community development efforts, the federal program has been highly politicized—and is likely to remain so. Private and nonprofit programs can choose to maintain a low profile (as the LISC demonstration program did in its early stages) and generally do not attract the same scrutiny as public programs, in which accountability for public dollars is always a central issue. The EZ/EC chose to be highly visible to attract the best possible pool of applicants (and to serve political purposes); in so doing, it raised expectations among prospective participants beyond those in the LISC demonstration and most community development initiatives. Finally, the Title XX grants—small relative to zones in big cities, but large relative to the sums generally available for community development—guaranteed that the programs would have high visibility and would attract controversy. Whether they also attract and sustain meaningful resident participation—and the commitment and capacity that could follow from it—remains to be seen.

Community Organizing as a Community Development Strategy

Community organizing was once commonly viewed as a core element of community development. However, as CDCs became more pervasive and as they became more focused on real estate development, the advocacy activities of community development became less prominent. Declining political and public support, as well as limited funding for organizing, was part of the cause. The other part was the difficulty of having a single organization try to simultaneously lead confrontational protests against financial institutions and city hall and then seek funding (often from those same sources) for development purposes. This led many (e.g., Shiffman and Motley 1989) to criticize the CDC movement for abandoning its core purpose and to conclude that community organizing and community development were incompatible.

In fact, viewing community development in social capital and networking terms facilitates an understanding of the role of community organizing in community development—understood broadly as the restructuring of political, economic, and social relationships to permit disinvested neighborhoods to produce a higher quality of life for residents, rather than simply as physical rebuilding. As Hornburg and Lang (1998) noted, social capital provides a policy language for expressing ideas that community organizers have long held as common knowledge.

Confrontational Organizing

Community organizing has often been confrontational and associated with struggles for political empowerment. Traditional organizing efforts, such as those undertaken by the IAF, the Association of Community Organizations for Reform Now (ACORN), and the black power movement, were based on political action. The principle underlying these efforts is that certain racial and socioeconomic groups have been systematically discriminated against and that confrontation with vested interests is necessary to overcome discrimination and increase economic and social opportunities.

The focus of traditional organizing is on bonding capital—the strengthening of internal ties among people and organizations sharing similar values and interests. Working mainly through established organizational networks, primarily churches, these efforts mobilize residents for actions that confront powerful people and institutions in an effort to get them to behave differently. Their goal is to change "the system."

In traditional conflict organizing, there is little value and energy placed in the development of weak ties or bridging social capital, as strong ties are thought to be sufficient to empower actors and effect change. Some conflictual organizers in practice explicitly reject developing weak ties with those in power for fear of having group members "co-opted" or feeling "disempowered" when sharing responsibilities with people or organizations in an advantaged positioned in the system.

Some critics (following the logic of what happens when bonding social capital is emphasized without bridges) might view conflictual organizing negatively. Although in the short run, benefits may be derived by a disempowered group through conflict organizing, long-term conflictual organizing efforts could lead to social and political division, harm the ability of different groups to work together, limit the amount of funding and access to outside resources, and be detrimental to larger community and societal interests. This critique of a conflictual approach to community organizing was used by some to justify the alternative consensus-based approach used in the LISC demonstration. On the other hand, some conflict organizers (including several from the local sites whom we interviewed in our research) would argue that "pragmatic" community development efforts, such as pursued in the LISC demonstration and by many CDCs, end up serving narrow and relatively unimportant interests, including those of CDC professional staff, and not the general interest of the targeted community (M. Gittell et al. 1996). To avoid this, the LISC demonstration took special efforts to ensure that newly organized CDCs represented broad community interest (through the recruitment of diverse board members), were not heavily reliant on outsiders (including professional staff), and had bridges to others outside their own neighborhood (including other CDCs and members of the support community).

Consensus Organizing

Consensus organizing frames its goal very differently from conflictual organizing. The objectives are to develop neighborhood leadership, organize community-based and controlled organizations, and facilitate respectful and mutually beneficial relationships between neighborhood-based leaders and organizations and the larger metropolitan-area support community. In contrast to conflictual organizing, consensus organizing pays attention to the development of strong *and* weak ties, namely, both bonding—the nurturing of internal social capital (e.g., furthering trust and cooperation among neighborhood residents)—

and bridging (creating weak ties, i.e., working relationships) to those with resources, power, and influence. The logic is that there are benefits from parallel organizing—that is, organizing people in their neighborhoods and those with resources and power in support of those organizing in their neighborhoods—because one activity supports the other. Having people organized and ready to act in their neighborhoods gives a reason for the support community to organize, and having the support community organized and ready to help can be a motivating factor in organizing the residents of low-income neighborhoods. The blending of bonding and bridging social capital is thus at the core of consensus organizing.

To meet its objectives, consensus organizing differs from conflict-based organizing in two important ways. First, consensus organizing consciously seeks to cut across the lines of existing neighborhood interests to build new leadership and organizations, rather than work through existing channels. The goal is to create organizations and leaders that are broadly rooted and accepted in their community. Consensus organizing attempts to focus on issues that all (or most) parties can agree to—such as neighborhood housing improvement or redevelopment—that are not controversial and will result in minimal disagreement and conflict. In contrast, traditional organizing focuses on contested issues—those that will actively engage people because they feel strongly enough about them.

Second, rather than confrontation and legal action, the consensus organizing approach emphasizes self-help, with a priority on developing residents' capabilities and building positive linkages (bridges) to the larger metropolitan-area public, private, and nonprofit support community. The priorities are to help disadvantaged groups gain greater control over the development agenda of their communities and to fortify beneficial relationships with established metropolitan-area development actors. The twin tasks are to create community-based leaders and organizations that will initiate constructive activity on the community's behalf and to cultivate a system that will support that activity.

In contrast, conflict-based organizing is highly pessimistic about the potential for low-income residents to increase their control and influence, without confronting, challenging, and profoundly changing the system. In any relationship with those in power, the residents of low-income neighborhoods would be at a disadvantage and therefore would not be empowered by furthering weak ties, but instead be further marginalized by having to ask for resources and support.

The LISC national demonstration program is thus, among other things, a test of the efficacy of the consensus-based approach to organizing. Like the spate of foundation-initiated comprehensive community change initiatives under way across the country, it is an effort to create an environment that is richer in both bonds and bridges. However, it differs from other programs in placing explicit programmatic emphasis on building new "cross-cutting" relationships within

impede implementation of zones' strategic plans, and that competitions for federal funding from all relevant agencies would give priority to applications that would benefit EZs and ECs.

16. Whereas these four dimensions—long-term trusting relationships, shared vision, mutual interest, and financial nexus—capture the essence of partnerships supporting community-based housing production and related activities, we will need substantially more experience with EZ/EC program implementation before it becomes clear how well this construct applies to the EZ/EC case, which has greater variation and is significantly more complex.

17. Plans often call for private-sector funds to be invested directly, for example, in staffing training programs or in direct investments in plant and equipment in the zone, rather than passing through a public or nonprofit body.

Getting Off to a Good Start

Positioning the Program in the Field

Selecting sites, hiring staff to work in each locality, and targeting neighborhoods to organize in each site were the initial steps that lay the foundation for the rest of the program. In making these choices, particularly the choice of sites, LISC (i.e., Michael Eichler and Richard Manson) selected the contexts in which the program would be tried and influenced how the program would be perceived there. Most important, they identified a core of participants: private-sector supporters and a group of communities and their resident volunteers—the "raw material" with which the program would work. Site selection committed the program to working in distinct political, social, and economic environments to which a myriad of program elements, including staffing and neighborhood selection decisions, would have to be tailored.

Like most subsequent program elements, the activities described here—site selection, hiring of staff, and neighborhood targeting—were designed to promote multiple objectives. They included long-term goals, intermediate outcomes, and outcomes to strengthen the program's management and operation. Hence, examination of the choices within start-up activities, some of which appear to be rather simple, illustrates the program's underlying complexity and suggests the program's approach to managing risk.

Objectives and Strategy

LISC's overarching objective was to expand the national community development industry into localities where it previously did not exist. The demonstration program was a test of the general efficacy and applicability of the consensus organizing approach (as described in Chapter 2) in furthering that objective, and that fact influenced the selection of sites and neighborhoods. The program sought to show that a broad range of localities and their low-income neighborhoods could benefit from a common approach to community development.[1]

Site Selection

The program's approach to demonstrating broad program efficacy was to be sure that each site (1) clearly needed community development assistance of the character the demonstration could provide and (2) also had the raw materials that would enable the program to succeed. In addition, to show the breadth of the program's possible applicability, sites were sought that, as a group, differed from one another in important respects (e.g., socioeconomic and demographic characteristics, size, and political context).

Assessing need was relatively straightforward. Site assessments relied on two primary indicators: the presence of neighborhoods with significant physical and economic problems and limited existing community development capacity, both at the metropolitan-area level and in neighborhoods.[2]

Assessing the likelihood of success was more complex and entailed three main elements. First, program success hinged on the presence of an interested and supportive private sector. Most obviously, private sector support would provide a strong likelihood that local funding for the program, required as part of the core LISC program, would be forthcoming. Just as important, however, sites were sought in which a group of highly regarded private sector leaders would give the program strong support beyond just financial assistance (for greater detail, see Chapter 6 on relationship building with the private sector) and thereby give it visibility and credibility.[3]

Second, program success critically depended on the presence of a pool of potential community volunteers. Assuming, as noted earlier, that the site had genuinely needy neighborhoods, sites were sought in which enough of those neighborhoods contained a critical mass of committed residents and other stakeholders who would be willing to volunteer and to support neighborhood-based CDCs.

Finally, because the program relied on consensus-style organizing, the likelihood of success was greater if there were no local groups likely to oppose or compete with the program. Some localities had such limited prior experience with community development that they literally presented a clean slate—the simplest program setting. Other localities presented more complex environments that included ineffective or nonrepresentative nonprofit development entities, discredited past efforts that had tarnished community development's image, or existing groups that relied on conflict-style organizing. In such places, the program looked for signs that any existing community organizing or development organizations were ones with which the consensus organizing effort could coexist. Together, these three criteria helped identify sites that would likely benefit from demonstration program efforts.

Staffing

The local development team in each site consisted of a coordinator and three community organizers. Eichler hired coordinators when it became clear that the local fund-raising target for the program was likely to be met. The coordinator, in turn, hired the community organizers.

The development teams were designed to operate independently of LISC local program directors. The staff was physically housed and the team officially administered by local organizations referred to as hosts. The intent was to give the development team a strong local connection that was viewed as critical, particularly at program start-up. Each host had a strong local reputation that often gave development team staff instant credibility.

The primary objective in assembling the staff was to ensure a team that could deliver the program at a high level of quality and integrity. This required, first and foremost, identifying individuals who could demonstrate that they understood and were strongly committed to the program's (1) core values and concerns—including diversity, honesty, responsiveness to local concerns and needs, and improvement of conditions and capacities of targeted populations—and (2) methods (i.e., the consensus organizing approach). To ensure this, Eichler chose not to recruit individuals with previous organizing experience, most of whom would have been trained in confrontational approaches to community organizing. He mainly sought people who had or could readily acquire the skills needed to implement the multiple program components and would be highly motivated by the core values and objectives of the program. Relatively low salaries (on average approximately $32,000 for coordinators and $19,000 for community organizers)[4] were established both to keep the program budget low (and more marketable to prospective local program sponsors) and to attract to the development team individuals who were "really" committed to the endeavor.

Another objective of staffing was to assemble a multitalented and diverse staff across the sites. One of the national program goals was to demonstrate that diverse, talented, and committed individuals could be attracted to the community development field—with the right program, a dynamic leader (Eichler), and a connection to a reputable national organization (LISC).

Neighborhood Selection

The primary strategic objective of the neighborhood selection process was to identify in each site a set of neighborhoods that would allow the program to demonstrate its value. As in the site selection process, the development team pursued this objective by identifying neighborhoods that clearly needed community development, but that also possessed the attributes the program needed to succeed. The focus on seeking conditions in which the program was likely to work was driven by both short- and long-term concerns. The immediate concerns were the program's short time frame and limited resources and the desire to show that good use had been made of those resources by producing programmatic progress in a relatively short period.[5] The long-term concern was that future support for community development in the demonstration sites (and perhaps in potential new sites) was contingent on "proven" success.

The principal criterion used in making the judgment about the likelihood of success was the prospect of recruiting a diverse, committed group of volunteers who would be interested in engaging in a new community development effort working with the development team. Following from this, the community organizers were instructed to assess the commitment of local residents to improving their neighborhoods, their leadership potential, and their willingness to volunteer for an organized group effort and work together.

Three other criteria were also considered: (1) the presence of potential institutional participants in the neighborhoods (e.g., churches, schools, hospitals, cultural organizations); (2) whether there were existing community development efforts reducing the pool of potential volunteers and raising the possibility of conflict over resources; and (3) whether there had been any recent negative experience with community development that would "jade" residents' outlook and make it difficult to recruit volunteers.

The neighborhood screening and selection process was designed to achieve multiple objectives, including establishing momentum for organizing efforts in the neighborhoods ultimately selected, generating broad support and enthusiasm for the program in neighborhoods and in the support community, collecting information, and fostering the professional development of newly hired community organizers.

The "motivational" aspects of the neighborhood selection process derived from the development team's use of potential for success as a selection criterion. (This is in deliberate contrast to many social programs and organizing efforts that use need and problems as the main criteria for receiving program funds and attention.) Consensus organizers assured prospective volunteers that if their neighborhood were chosen, significant community development assistance and resources would be forthcoming. They also emphasized that the proposed approach had been implemented successfully in other areas, so if their community was chosen and neighborhood residents did "the right things," they had a good likelihood of success. The rationale for emphasizing potential for success was that, unlike needs and problems, it could instill confidence in neighborhood residents and motivate constructive activity, including volunteerism.[6] It could also stimulate healthy competition among residents of different neighborhoods who hoped they would be chosen, a further motivating influence.[7]

Neighborhood selection was also intended to widen the program's base of support (in individual sites and nationally). This was to be accomplished through *neighborhood coverage,* that is, the development team tried to ensure the selection of a group of neighborhoods that represented an important and diverse range of needs and demands and that were located in different local jurisdictions or city-council districts. Increased support for development team efforts was meant to be further encouraged by making clear to potential supporters that the program was targeting neighborhoods where success was likely; this approach would have particular appeal to private sector supporters. Finally, positive media attention for the "chosen" neighborhoods and their volunteers was also intended to build broad support for the program.

Neighborhood selection provided community organizers with an important, focused responsibility immediately after their training and helped to initiate positive ties between the community organizers and "their" neighborhoods. Community organizers' evaluations involved extensive fieldwork and interviewing, providing a clear structure and a distinct reason for gaining knowledge about the neighborhoods they would eventually organize in and establishing connections with neighborhood residents.

In addition to considering the characteristics of individual neighborhoods, the development teams sought neighborhoods that, as a group, gave a strong presence to communities of color and represented a range of problems and related program risks. The emphasis on communities of color was driven by the fact that these communities have commonly been plagued by disinvestment and have been politically marginalized (Oliver and Shapiro 1995); thus, to demonstrate its value, the national program had to demonstrate that it could be effective in assisting these neighborhoods. Variations in physical, economic, and social conditions of the neighborhoods had the benefit of allowing the program to include neighborhoods where there was a real possibility of failure, without jeopardizing the program as a whole. The program would then be less vulnerable to criticisms of "creaming," that is, helping the neighborhoods that least needed assistance.

Activities to
Implement the Strategy

Site selection began when one or more leaders from a locality contacted LISC to express interest in the program. If preliminary conversations went well, LISC challenged the locality to raise a modest sum ($10,000) to cover the costs of a full site assessment. These funds, once raised, supported at least two site visits by senior LISC staff, with Michael Eichler and Richard Manson playing the lead roles. These visits included interviews with a broad cross-section of more than 75 local corporate and civic leaders, including individuals knowledgeable about the area's neighborhoods and significant numbers of stakeholders in those neighborhoods.

A recommendation that the site was a promising place for the consensus organizing program led LISC to challenge would-be sponsors at the site to demonstrate "local initiative" by raising a target sum, agreed on by LISC and the local sponsors, to seed the local LISC grant and loan pool and to cover the local costs of the demonstration program (including development team staff). Eichler and other senior LISC staff supported the fund-raising by meeting individually with prospective donors and appearing at fund-raising events designed to draw positive media coverage.

As with other program activities, site assessment served multiple purposes. By the time fund-raising was complete, Eichler had accumulated a substantial network of contacts, including the names of people who might be good community organizers or neighborhood volunteers and a list of neighborhoods those contacts thought might be good places for the development team to work in. He also had an understanding of the priorities of the different public, private, and nonprofit leaders he interviewed—including an idea of which neighborhoods (or types of neighborhoods) would have to be included to get the desired level of support for development team efforts. This information later became raw material for the neighborhood selection process.

When it was reasonably clear that the fund-raising target was likely to be met, Eichler hired and trained a local coordinator to manage the program in the new location, and the program was formally under way.[8] The coordinators, once hired and on site, began to build a network of contacts, to become familiar with potential target neighborhoods, and to recruit and hire three community organizers.

Three sets of activities were used to lay the foundation for local support of program staff: broad outreach to identify candidates, careful screening of the candidate pool, and involvement of a diverse and representative committee of local residents in interviewing the finalists for the staff positions. Eichler asked many people about possible candidates for the staff positions during the process of screening sites and fund-raising. He also assembled a group of between six and eight people to serve on coordinator interview committees; these were individuals who he believed genuinely understood the program goals and approach, who were members of constituencies whose support the program would need, and whose judgment would be respected by others in the community.

The coordinators began their work by meeting with various stakeholders and with prospective organizers identified by Eichler during site assessment. The coordinators used the need to identify a strong community organizer candidate pool as a rationale to meet with people and to expand their networks of local contacts.

From a long list of candidates, the organizer pools were narrowed. After screening and extensive interviews with candidates, each coordinator assembled another interview committee, including some members of the initial coordinator interview committees and some new individuals with whom they wanted to establish strong program ties.

By the time the organizers were hired, the coordinators had identified a shortlist of between 9 and 12 potential target neighborhoods. These neighborhoods had been identified during site assessment and in conversations about organizer candidates. Prospective target neighborhoods were divided among the community organizers. The organizers' first task was to assess the need for, and potential efficacy of, a consensus organizing effort in each neighborhood.[9] After a three-month assessment period, the community organizers recommended the neighborhoods in which they thought the development team should work, and the local coordinators acted on their recommendations.

Neighborhood selection was highly competitive. The rewards were the commitment of development team resources, including technical assistance, in (1) organizing new CDCs, (2) building CDC board capacity, and (3) developing individual real estate development projects. In addition, the neighborhoods selected were all publicly acknowledged for their potential for successful community development, including the commitment of their residents, and were well positioned to draw on LISC's grant and loan pool.

Findings From the
Implementation Experience

Site Selection

The demonstration program experience indicates that the site selection approach used was effective in the service of multiple objectives. At both the site and neighborhood levels, the final program portfolio (the product of the program steps described previously) included places where the needs and resources of the communities met program criteria, but also exhibited enough variety to permit the program to assume a desired level of risk. In addition, the process that led to these choices succeeded in laying a strong foundation for subsequent program activities—by building support networks, positioning the program favorably with key local constituencies, and motivating program participants.

In choosing the initial three demonstration program sites, LISC considered 10 localities over a two-year period. Of these, five went through the formal assessment process: Palm Beach County, Little Rock, New Orleans, Phoenix, and Fort Worth. Phoenix was judged to be poorly suited to the development team approach. In addition to one large, nationally prominent CDC, the city had a solid core of about a half-dozen fledgling CDCs; they clearly needed capacity building assistance, but were past the point of needing the type of "start from scratch" help the demonstration program was designed to provide. Fort Worth ultimately elected not to participate.[10]

The chosen sites vary significantly in scale, extent, and severity of poverty, robustness of the local economy, housing market, political culture, climate of race relations, and prior community development experience. To some degree this is inevitable (every city has its distinctive features), but it was also a result of a deliberate decision to test the consensus organizing approach in varied circumstances to learn how widely applicable it is and what is required to make the approach work under different conditions. All three sites presented significant challenges.

Palm Beach County, best known for the very wealthy coastal community of Palm Beach, was the largest of the three demonstration sites and was experiencing the most rapid growth. Its population of approximately 863,500 had increased almost 50 percent since 1980, and the U.S. Census Bureau predicted that it would be the fastest growing large metropolitan area during the 1990s. The local economy was strong, and the county ranked ninth among all U.S. counties in number of building permits issued.

Underlying this rosy picture, however, were profound contrasts. With an 11 percent poverty rate, the county was characterized by considerable income disparities, and uneven development had led to increasing segregation. Much of the housing growth occurred in unincorporated subdivisions, and property values were steadily rising beyond the reach of many lower-income households. Minorities, comprising only 13 percent of the population, became increasingly concentrated within the county's larger, older cities.[11] Although an Affordable Housing Task Force found that the county would need an additional 6,000 affordable units per year to meet the needs of low- and moderate-income families, local government regulations to slow development made such production difficult. Some county-wide efforts were initiated to address these problems. For example, banks formed an affordable housing consortium, and in an effort to increase minority voting influence, the County Commission (the county's most powerful political force) was altered in 1988 so that the seven members represented single districts. Several neighborhood associations and church-based organizations were active in the county's communities, but none undertook real estate development.

Palm Beach County's intercoastal communities had the highest concentrations of poverty and of minority residents in the county outside the Glades (a predominantly rural area discussed later). In these communities, there had been a paucity of community development effort. Residents suffered from lack of infrastructure, affordable housing, and community services. They also suffered from a history of racism and neglect. Finally, they lacked effective community-based organization and leadership that could initiate community development effort.

In contrast to Palm Beach County, Pulaski County, located in central Arkansas, is a relatively small urban area with only 350,000 residents, many of whom live in Little Rock, the state capital, and North Little Rock, located across the river from Little Rock. The county had been growing, but much more slowly than Palm Beach, and state government remained the largest employer. Older neighborhoods in both inner cities had experienced substantial population declines as more affluent, white residents moved to the suburbs of these cities. The integration efforts and civil rights activism of the 1960s, including the federally enforced integration of Central High School in Little Rock, spurred this movement. Local planners predicted that gradual population losses would continue.

African Americans composed 34 percent of Little Rock's population in 1990, versus 27 percent countywide. As the suburbs saw development of new houses and businesses, older inner-city neighborhoods experienced significant housing decay and increasing crime rates. The 1990 census indicated a county poverty rate of 14 percent and suggested that, on average, white residents had more than twice the per capita income of African American residents.

Both Little Rock and North Little Rock had numerous low-income, predominantly African American neighborhoods bordering their downtown areas. These communities (similar to the intercoastal communities in Palm Beach County) had suffered from an extended period of neglect and a history of racism. Also similar to the intercoastal communities in Palm Beach County, there was very limited community-based development capacity and control. In contrast to the other two sites, however, Pulaski County had some existing organizations that operated citywide, had been active, and had support and credibility within significant circles. A coalition of neighborhood associations had played a prominent role in blocking a proposed development project in downtown Little Rock just prior to the start-up of the LISC program. In addition, Little Rock is the home of the Association of Community Organizations for Reform Now

	Gender	Race/ Ethnicity	Local/ Nonlocal	Education
Palm Beach County				
Mary Ohmer (Coordinator)	Female	W	Nonlocal	Master's: Public & Internal Affairs; Social Work
Navara Peterson	Male	AA	Local	Community college
Joanna Tarr	Female	W	Nonlocal	Master's: Public Management & Policy
Lorenzo Young	Male	AA	Local	BS: Management
Little Rock				
Richard Barrera (Coordinator)	Male	H	Nonlocal	Master's: Public Policy
Felicia Cook	Female	AA	Local	High school + some business college
Carl Dokes	Male	AA	Local	BA; 2 years toward MA
Willie Jones	Male	AA	Local	BA: Political Science
New Orleans				
Reginald Harley (Coordinator)	Male	AA	Local	BA: Economics & Business
Marian Brownson	Female	W	Local	BA: Psychology (almost complete)
Eric Johnson	Male	AA	Local	Master's: Public Administration
Karen Johnson	Female	AA	Local	BA: Business Administration

Figure 4.1. Characteristics of Development Team Staff
NOTE: W = White; AA = African American; H = Hispanic.

Staff

Despite the time commitment and dedication required and the comparatively low salaries paid, the program was able to attract individuals with the right talents and commitment to local development team staff positions. The program staff was diverse, both racially and socioeconomically (see Figure 4.1 for selected characteristics of the local program staff).[13] One of the three local coordinators and seven of the original nine community organizers were African American, one of the coordinators and four of the organizers were women, and one of the coordinators was Hispanic.

Local supporters viewed the staff favorably from the outset. Their affiliation with LISC gave staff the benefit of the organization's established reputation for quality. They had the additional advantage that they were approved by Eichler, in whom supporters rapidly developed great confidence. At the same time, the members of the interview committees were both diverse and well regarded in their communities. The fact that the individuals they recommended were hired positioned the program as being responsive to local views and needs.[14] The experience with staffing is a vivid example of how relatively simple program activities served multiple purposes.

The strong representation of people of color was critical to building local credibility for the program, especially in the target neighborhoods, because it helped to undercut the distrust of some in the African American community who saw LISC as a "white, New York organization."[15] It also gave the program credibility for serving as an example (model) of the inclusive behavior it expected of others.

The task of identifying a mix of candidates that would ultimately yield a capable, committed, and diverse staff was more labor intensive than most search processes. It required tapping into a wide variety of formal and informal networks, both locally and nationally. Interviews with sources of names of prospective job candidates were far more numerous than interviews with actual candidates. Effective screening of the candidate pools required Eichler to interview numerous candidates to identify prospects strong enough to be seen by the interview committee.

The coordinator role required a unique combination of qualities to perform successfully with limited supervision. Therefore, extensive reconnaissance—typically beginning well in advance of when the position needed to be filled—was necessary. Local coordinators were recruited nationally and locally. Even with a national recruitment horizon, the pool of suitable candidates was small, and the number of local candidates was particularly limited. Eichler was able to secure the services of the candidates he felt were most able, and they were all clearly capable professionals. Nevertheless, the difficulties the program encountered—even with a carefully hand-picked staff, testify to the challenging nature of the job and the potential problems this could create in efforts to replicate the program.

For the positions in Palm Beach County and Little Rock, Eichler focused his efforts nationally, mainly among his own network in the community development field. He hired Mary Ohmer as the coordinator in Palm Beach. She knew Eichler from Pittsburgh where they both received their master's degrees in social work at the University of Pittsburgh. She was familiar with Eichler's efforts in

Pittsburgh and the Mon Valley and highly motivated to work with him on the national demonstration. Ohmer had worked in community development in Pittsburgh after graduation. She was young (early 30s), energetic, and self-confident, enthusiastically moving from her home in Pennsylvania to Florida.

For the coordinator position in Little Rock, Eichler hired Richard Barrera. Barrera was the youngest (24) and least experienced of the coordinators hired. He had first come into contact with Eichler when Eichler was a guest speaker at Ross Gittell's community revitalization class at the Kennedy School of Government in the spring of 1990 and Barrera was a student in the class. Barrera was intrigued by Eichler's presentation of the consensus organizing approach to community development and arranged to meet separately with Eichler after class. After Barrera received his master's at the Kennedy School, he attended Law School at the University of California at Berkeley. He left law school after his first year and called his former professor and informed him that he "wanted to work for Mike [Eichler]." Gittell told Eichler, who then contacted Barrera directly. This was before funds for the national demonstration had been secured so Eichler told Barrera that he would keep in touch. In the meantime, Barrera went to work for the city of Oakland on community development issues.[16] When local program funds seemed ensured in Little Rock, Eichler invited Barrera to visit the city for an interview. Barrera was hired, moved to the city with his wife, and began work in advance of the official start date of the demonstration program effort.

In New Orleans, the local context (as described earlier) suggested the importance of hiring a coordinator from the city and an African American. After a long search in which few qualified candidates emerged, Eichler hired Reggie Harley. The weak pool of candidates was not surprising given the history of community development in the city. Even under these circumstances, Eichler was able to identify a coordinator in whom he had confidence. Harley had been living in New Orleans for more than 10 years and had previous economic development experience in the city as well as international development experience with the Peace Corps. He was in his late 30s and was married with two young children. Harley first heard about LISC coming to New Orleans and Eichler's interest in hiring a coordinator from a friend at his church who was a program officer at the Greater New Orleans Foundation. He then contacted Eichler, who was most impressed by his work in the African American community and his reputation in the city.

In contrast to the coordinator position, the pool of community organizer candidates for all of the sites was more than adequate for the needs of the development team. The pools were fairly deep and diverse in talent and socioeconomic characteristics. In recruiting community organizers, the most important qualities needed for the job—deep commitment, an ability to relate well to people, and common sense—were hard to detect from a resume. Instead, personal recommendations and multiple interactions with candidates were used to explore and probe an individual's qualifications.

Consistent with the stated intent of fielding a consensus-based organizing effort, none of the staff hired for the program had prior community organizing training or experience. Further, because the development team staff's role was organizing, familiarity with the real estate development process was not required.[17]

The orientation and initial training of the community organizers focused heavily on volunteer identification and recruitment. This made sense because (a) the primary criterion for selecting neighborhoods (and the key ingredient in ultimate success in organizing them) was the ability to recruit a diverse group of good volunteers and (b) the organizers, by design, had no prior organizing experience. Eichler highlighted in the training six *neighborhood networks*—homeowners, renters, businesspeople, religious organizations, social service providers, and prominent neighborhood institutions (e.g., hospitals and schools)—that the organizers were expected to assess for their commitment to the community and their likelihood of volunteering to join a collective community development effort. Community organizers learned to interview in each of the six networks, including how to assess leadership potential, attitudes toward community development, and willingness to volunteer.[18]

The neighborhood selection process was a critical part of the organizers' professional development. Conducting the neighborhood assessments was their primary initial activity. The task provided focus and specificity to their first three months on the job. The overall objective and their individual responsibilities were clear, they had been trained to do the tasks required to meet their responsibilities, and they had a deadline.

Neighborhood Selection

During the site selection process, Eichler determined that each of the sites contained enough neighborhoods with the two most important requisites—demonstrated need for community development to give the program credibility and a pool of motivated prospective volunteers with leadership potential—to permit the program to achieve the desired scale. In his interviews with local leaders, he solicited suggestions about which neighborhoods should be targeted for development team efforts. Again, interviews served multiple objectives. Locally generated recommendations helped Eichler develop an understanding of the priorities of the different public, private, and nonprofit leaders he interviewed. This information then helped the development team address the *coverage* issue, namely, which neighborhoods (or types of neighborhoods) to include to get the desired level of support for development team efforts.

The local coordinators, often in consultation with Eichler, ruled out some prospective neighborhoods before the formal neighborhood assessments began. Some communities faced problems that were so severe or unique that the program was an inadequate remedy. For example, Eichler and Mary Ohmer, the local coordinator in Palm Beach County, rejected the idea of working in the Glades communities in the western part of the county. The Glades were physically isolated from the rest of the county and radically different in character from the poor neighborhoods near the intercoastal waterway; their development problems were much more severe, more rural in nature, and more related to the fundamental economic dependency relations of an economy built on sugar cane production. In New Orleans, the local coordinator and Eichler ruled out early the option of working in communities dominated by large, distressed public housing developments. In both instances, it seemed clear that the program strategy and resources were inadequate to address the problems at hand, and that LISC would risk wasting both program resources and residents' time and hope. At the other end of the spectrum, a few neighborhoods were eliminated from active consideration because simple reconnaissance showed them to be comparatively well off.

The neighborhoods chosen for participation in the program were clearly neighborhoods with significant community development needs (see Tables 4.1, 4.2, and 4.3). All had average household income at least 25 percent below their county's average, all had problems with their housing stock (creating the basis for an initial set of housing development projects), and all lacked the organizational resources to undertake development activities. Most of the neighborhoods were predominantly African American, but each site had at least one neighborhood that was racially and ethnically mixed. As a group, the chosen neighborhoods in each site had lower income (and hence presumably greater need) than those assessed but not chosen.

The neighborhoods chosen at each site were geographically, and hence politically, dispersed—located in five cities and an unincorporated area in Palm Beach County, two cities and a partially incorporated neighborhood in Little Rock, and in different council districts in all three sites. Although the neighborhoods shared many problems, they also provided different contexts for community development, as the discussions in the next chapter will illuminate.

The neighborhood selection process illustrates the way in which the program tried to manage three types of risk: (1) organizing new CDCs where existing groups were potential competitors, (2) working in neighborhoods in which the public sector might be unlikely to invest, and (3) working through existing groups.

Attempting consensus organizing in communities that already had an organizing effort under way immediately raised a red flag. The consensus organizing program represents potential competition for a program that relies heavily on conflict. Two such efforts were active in the chosen sites: ACORN was founded

(text continued on p. 76)

TABLE 4.1 Characteristics of Neighborhoods—Palm Beach County

| | | | | | Palm Beach County, Florida 1990 Census Data | | | | |
	Median Household Income ($US)	Unemployment Rate (Percent)	Labor Force Participation Rate (Percent)	Percentage Households With Public Assistance	Percentage Minority Population	Percentage Persons Age 25+ With No Diploma	Vacancy Rate (All Units; Percent)	Percentage Renter Occupied	Total Population
County	32,524	5	43	11	13	21	9	25	863,518
Chosen Neighborhoods									
Boynton Beach	22,600	10	63	51	86	40	13	35	8,442
Del Ray Beach	22,141	8	69	53	91	64	15	49	10,126
Lake Worth	20,540	21	65	16	47	38	21	52	7,278
Limestone Creek*	54,132	17	65	4	7	22	9	16	1,944
NW Riviera Beach	17,590	10	69	41	99	27	12	42	4,643
Pleasant City	14,845	7	61	55	83	55	24	84	2,084
Neighborhoods Not Chosen									
Grandview/Flamingo	24,030	14	71	7	47	38	13	50	3,506
Lake Park	29,167	19	61	10	24	23	13	42	6,704
Northwood	24,159	11	65	18	58	34	15	45	13,704

SOURCE: U.S. Bureau of the Census. 1992.
NOTE: *Limestone Creek is a very small neighborhood; census data describing it are aggregated by the census with data from surrounding wealthy areas.

TABLE 4.2 Characteristics of Neighborhoods—Little Rock

Little Rock, Arkansas
1990 Census Data

	Median Household Income ($US)	Unemployment Rate (Percent)	Labor Force Participation Rate (Percent)	Percentage Households With Public Assistance	Percentage Minority Population	Percentage Persons Age 25+ With No Diploma	Vacancy Rate (All Units; Percent)	Percentage Renter Occupied	Total Population
Pulaski County	26,883	5	66	6	27	14	9	40	349,660
Chosen Neighborhoods									
Argenta	11,780	3	49	17	21	8	6	64	4,913
Central	14,147	10	56	16	92	29	20	49	7,452
College Station	13,215	10	44	19	91	24	6	31	1,415
Downtown	13,502	9	56	15	75	19	24	63	8,067
McClellan	23,848	6	72	6	45	19	12	47	11,301
South Little Rock	14,651	11	57	20	98	27	12	38	5,576
Neighborhoods Not Chosen									
Baring Cross	15,506	13	58	11	28	26	14	42	1,458
Dixie	14,087	4	56	25	100	39	10	21	608
Pankey	45,237	3	73	1	7	4	5	26	14,674

SOURCE: U.S. Bureau of the Census, 1992.

TABLE 4.3 Characteristics of Neighborhoods—New Orleans

New Orleans, Louisiana
1990 Census Data

	Median Household Income ($US)	Unemployment Rate (Percent)	Labor Force Participation Rate (Percent)	Percentage Households With Public Assistance	Percentage Minority Population	Percentage Persons Age 25+ With No Diploma	Vacancy Rate (All Units; Percent)	Percentage Renter Occupied	Total Population
Orleans Parish	18,477	13	67	14	65	22	17	56	496,938
Chosen Neighborhoods									
Algiers Crescent	13,506	18	55	18	74	27	28	59	2,045
Bacatown	8,575	32	47	30	97	32	18	80	2,551
Gert Town	10,519	14	43	21	97	37	18	76	2,327
Holy Cross	14,193	22	49	22	80	27	19	61	3,051
Mid-City	15,960	14	52	13	59	24	20	73	2,786
New Life	13,478	19	53	22	99	29	16	40	3,241
New Vision	8,466	24	45	34	84	28	27	88	2,028
Neighborhoods Not Chosen									
Bywater	11,420	16	55	20	69	31	23	69	2,674
Irish Channel	12,892	23	50	32	76	24	27	79	1,753
Pensiontown	17,623	11	53	8	56	12	16	65	1,625
Seventh Ward	11,668	19	53	25	94	32	22	67	2,223

SOURCE: U.S. Bureau of the Census, 1992.

in Little Rock and had a local chapter that had been active long before the LISC program began. All Churches Together (ACT) had been active in New Orleans for several years and had fielded a series of "events" (public meetings) confronting the mayor. In Little Rock, Eichler determined during his site assessment that the program could work even with potential resistance and contention with ACORN. He felt that ACORN lacked the broad support and community development agenda that could deter LISC's efforts in the city. After program activities began in Little Rock, the local coordinator tried to reach out to ACORN and develop a working relationship but was largely unsuccessful. ACORN was never ideologically comfortable working with LISC and was highly doubtful about the potential efficacy of the consensus organizing approach, which contrasted with their own confrontational tactics. ACORN at times tried to undermine LISC's efforts (e.g., claiming that LISC groups were "selling out to corporate interests"), but was largely unsuccessful.

In New Orleans, ACT was particularly strong in Bacatown, where St. Peter Clavin Church was their local base of operations. Eichler and the local coordinator met with senior staff of the organization and concluded that the group did not pose a significant problem. ACT, in the opinion of Eichler and Harley, had little interest in pursuing a neighborhood development agenda. ACT did eventually influence the choice of neighborhoods in New Orleans, however. The development team had decided not to work in Central City because the organizer lacked confidence that he could find a large and diverse enough group of volunteers. After ACT successfully organized a well-attended event in the community, the development team agreed to work there. As it turned out (as discussed in the next chapter), New Vision (the Central City CDC) was at first weak—the board was neither large nor strong enough, with the result that work progressed slowly during most of the demonstration period. However, with new church-based leadership from Father De Voe, by the summer of 1997 the group was one of the stronger ones in New Orleans according to the local LISC program director. Another one of the stronger CDCs in New Orleans was in Bacatown, ACT's home community.

Two of the neighborhoods targeted by the program were places in which it appeared that the public sector would be reluctant to invest, thereby making it more difficult to finance projects that required subsidy dollars (which virtually all affordable housing developments do). One, Limestone Creek, is a small, isolated, low-income, African American community in the unincorporated northern part of Palm Beach County. It had limited infrastructure—no public water or sewers, only two paved roads—even though the communities that surrounded it (all of which are predominantly white) have roads and utilities. The severity and distinctiveness of Limestone Creek's problems raised the possibility that identifying and implementing a project (particularly one that was similar enough to projects chosen in other neighborhoods that LISC could support it effectively) would be difficult. Moreover, long-standing neglect and mistreatment of the neighborhood had made residents skeptical that their situation could be changed. These difficulties were compounded by the neighborhood's small size (well under 1,000 people), which seriously limited the size of the volunteer pool and thus raised the additional risk that it might prove impossible to build a strong CDC board. A directly parallel set of issues surrounded College Station, with the wrinkle that the City of Little Rock was explicit in saying they would not put city funds into projects in the part of the neighborhood that was unincorporated.

In spite of the problems, the development team elected to work in these two neighborhoods for several reasons. A demonstration that its approach would work in such difficult settings would be powerful evidence of the usefulness of the consensus organizing approach. Locally, each of these neighborhoods was commonly recognized as being seriously disadvantaged, so the program would in any event get credit for being willing to tackle hard problems. In the worst case, if the program did not successfully organize CDCs in Limestone Creek and College Station (but succeeded in the other neighborhoods), the severity of the community's problems, rather than the program's quality, would be seen as the reason and LISC would still be able to point to good outcomes in five of the six neighborhoods targeted—a very good record. Finally, and perhaps most important, the organizers assigned to assess the neighborhoods had each met a core of committed volunteers and believed strongly that the program could be successful there. Trusting the organizers' judgments and allowing them to take a risk by assuming responsibility for a difficult job reinforced central themes in their training and motivated them to make the program work in these neighborhoods.

Finally, although the consensus organizing approach works most readily with a clean slate, the program decided to work in a number of neighborhoods that had existing neighborhood-improvement groups (but not CDCs). These neighborhoods included Pleasant City and Lake Worth in Palm Beach County, South Little Rock, and Holy Cross and Gert Town in New Orleans. Program experience in working with preexisting organizations (discussed in the next chapter) illustrates the of difficulties of this approach. For example, Gert Town was the only targeted neighborhood in which a CDC did not receive pre-development funding.

Several factors drove the decision to work in these neighborhoods with existing groups, despite the expected pitfalls. They were all neighborhoods where Eichler and the staff felt the program should be working. They clearly had the two key ingredients: They needed help, and the organizers had found good neighborhood people to work with. In fact, the existing groups included exactly the types of people the organizers would want to recruit. However, a consensus organizing effort would not want to compete. Even if the new CDC and the existing group reached an implicit accommodation (as happened with ACT in New Orleans), it would be likely that they would ultimately compete for resources—money, volunteer energy, and political clout—and the CDC organized by the development team would then be undercutting a legitimate community effort.[19] In each case, the local coordinator (sometimes with Eichler) approached the group during the neighborhood assessment period to see if they would be willing to (1) adopt the program's agenda and (2) broaden their board to be representative of the entire community. Agreement by the group was taken as an initial sign that if the group actually met these two conditions, any problems that arose could probably be worked through.[20]

Each neighborhood on the short list that was not chosen had at least one characteristic that made it less attractive in terms of the key selection criteria. For example, in Palm Beach County, residents of Flamingo Park felt they were making progress in improving their community without a CDC and did not need one. Residents of Dixie (in Little Rock) shared similar sentiments—that the community could do very well without a CDC. Grandview Heights in Palm Beach County had an existing informal community group that was not interested in LISC's agenda; they were active in combating housing abandonment and drug-related crime and sought to attract more middle-class families to the neighborhood. In Irish Channel and Northwood in New Orleans, the organizers found deep divisions among residents, rooted in past conflict, which suggested that organizing a group that would be representative of the entire community would be very difficult.

The process of neighborhood selection also provided a springboard for the community organizers' organizing efforts after neighborhoods were selected. Each of the three community organizers in each site was assigned three neighborhoods to assess. The program's objective was to choose six of the neighborhoods on the shortlist in each site.[21] Through their early interviews, the organizers identified potential board members for the CDCs they would organize in the target neighborhoods. The depth and breadth of volunteer commitment served as the single most important factor in shaping their recommendations about whether a given neighborhood should be chosen. In all of the targeted neighborhoods, the organizers had elicited expressions of interest from a nucleus of volunteers. In addition, they saw signs that a wider group of residents might be receptive. For example, in Palm Beach County, Pleasant City and Lake Worth, there were preexisting community-based organizations interested in development.[22] Delray Beach supported an unusually high level of ongoing volunteer activity. Church members and small business owners in Boynton Beach voiced strong interest in working with the development team to strengthen their neighborhood, while volunteers in Northwest Riviera Beach were in the early stages of forming a group dedicated to keeping neighborhood youth away from drugs. In Little Rock, there was already a strong core of residents engaged in South Little Rock and downtown. In New Orleans, there were active residents in Holy Cross who were committed to objectives similar to those of the LISC program. In each site, the coordinator accepted the organizers' recommendations in their entirety, reinforcing the organizers' growing confidence in their ability to do their job.

Program staff organized a formal media event to announce the neighborhoods that had been chosen to participate and acknowledge the prospective volunteers in each community. Like many aspects of the program, these events had multiple purposes, each of which appeared to be accomplished. This included gaining publicity for LISC's efforts, solidifying and enhancing existing support by demonstrating that work had already been done and had been done on time (or earlier than expected), and giving momentum and confidence to individual and neighborhood efforts. Each of the media events announcing neighborhood selection was held at a prominent location, got good media coverage, and was well attended by both potential supporters and prospective volunteers. For example, in Palm Beach County, more than 100 people—including 34 residents from targeted neighborhoods—attended the announcement of the six chosen neighborhoods, and the program received positive coverage from two local papers and a local television station.

Lessons

The program did not "cream," that is, select "easy" sites and neighborhoods. The site and neighborhood selection process resulted in the choice of areas genuinely in need of community development. The program did, however, seek to identify sites and neighborhoods that were ripe for consensus organizing in the sense that they had strong volunteer energy and commitment, a paucity of existing effort, and a receptive and potentially forthcoming support community. This is another way of saying that the development team looked for program requisites, such as resident and support-community commitment to community development, that it could develop and leverage. As the following chapter makes clear, the program was, in fact, very successful in identifying sites and neighborhoods in which the staff could organize community organizations and support for these organizations.

The conditions used to screen sites can be found in a variety of locations, suggesting that there is potential to deploy the approach in other localities. The screens clearly rule out some areas, for example, those where the local community of prospective donors has inadequate enthusiasm for this approach (like Fort Worth), and those where there are already established community-based organizations (like Phoenix). However, the criteria employed do not limit the field to a small number of places.

Differences in local context influence program strategy and choices from the outset, for example, the choice of a coordinator (local versus nonlocal, race), of people selected for the interview committee (to build legitimacy for the coordinator choice), and of local sponsor (see Chapter 6 for more detailed discussion). Engaging a broad spectrum of people in this process gets their buy-in, adds credibility, and projects the impression that program support is broad. The diversity of the staff in each site helped build credibility, particularly in skeptical minority communities that have been previously disappointed by community development, particularly efforts initiated by outsiders.

Despite its obvious costs, the labor-intensive staff-recruitment process appears to be critical to the program's success. It is unlikely that a more conventional recruitment process could identify a diverse group with the right mix of commitment and skills. This is especially true for the local coordinator position, which (as we highlight in subsequent chapters) is pivotal to the program's quality and requires an unusual and demanding combination of talents.

Neighborhood selection, like site selection and staff recruitment, cannot rely on a "cookie-cutter" process. The criteria applied in identifying the potential to form a good CDC are more time-consuming and more subtle to apply than those used in assessing the capacity of existing CDCs. The outreach and networking process needed to identify a good mix of committed volunteers is labor-intensive, and the personal qualities the community organizers must assess, particularly personal integrity, commitment to the community, and an understanding of the program's approach, require time to elicit and judgment to discern.

Neighborhood selection is a multipurpose process. Although it appears simple, neighborhood selection must be thought-out, positioned, and followed up with consideration of many threads that are parts of different strategies. An important test of whether the neighborhood selection process was successful is whether the team was able to organize strong CDCs in those neighborhoods. The next chapter discusses the organizing of new CDCs in greater detail. Despite the problems noted there, the development team's significant progress in organizing and strengthening these groups suggests that the neighborhood selection process laid a good foundation for its work.

Notes

1. Finding commonality among diverse neighborhoods in each site also had benefits. With regard to program implementation, it allowed for more efficient effort during training, organizing, and project activity. It also created the potential for future collaborative effort among neighborhoods to garner increased resources from the public and private sector support communities.

2. Ideally, each site would have enough such neighborhoods to give some latitude for choice.

3. The private sector was emphasized because of its central role in the program design. However, the site visits did assess the likely receptiveness of the local public sector. Sites were not excluded if the public sector was unreceptive, but a positive public-sector response was considered a definite plus, and strong public-sector opposition to the program (a situation not encountered in any of the sites assessed) would presumably have generated reservations.

4. These salaries varied by site and were adjusted for labor market conditions. The highest salaries were paid in Palm Beach County, reflective of relatively high living costs in the area.

5. The two-year initial period, although putting pressure on the development team to show results, had the benefit of serving as a motivator for both development team staff and community volunteers. However, it proved to be too brief a period both to organize new CDCs and to have them make significant progress on a real estate project. In New Orleans, the program raised funds to cover an initial three-year period.

6. Neighborhoods not originally selected also had an incentive to "get their act together." This benefit of neighborhood selection had been realized in the Mon Valley where, after the selection of the initial six successful target areas, there are now more than double that number of CDCs as part of the Mon Valley Initiative (R. Gittell 1992).

7. Competition is healthy when it motivates volunteer commitment, helps to bring together preexisting efforts, and works toward the increase of resources available to community development.

8. Although called the *local coordinator,* the site manager was a local resident only in New Orleans. The coordinators in Palm Beach County and Little Rock moved to those places to take this job. In each case, Eichler trained the coordinator in a series of extended and intensive one-on-one sessions.

9. Neighborhood selection involves identification of general areas in which to focus development team organizing activities. The final decision about the boundaries of targeted neighborhoods was made by the CDCs when they prepared their articles of incorporation.

10. After the selection of the third and final demonstration site (New Orleans) in December 1991, LISC continued to receive expressions of interest in the program. More than half a dozen cities received at least some active consideration, but only four went through the full assessment process. Two of these, Baton Rouge and Las Vegas, were added to the program in 1993; they are not included in this assessment.

11. The average price of a home in Palm Beach County increased 79 percent from 1980 to 1987, and by 1990, 21 (more than half) of the county's municipalities had fewer than 2 percent African American residents. In that year, the federal government threatened to withhold educational aid because of the county's high and increasing level of segregation.

12. In Palm Beach County and Pulaski County, only 8 and 9 percent of households, respectively, were in this category.

13. Staff had to demonstrate their commitment to the program by agreeing to stay for the full period of the demonstration. Turnover among the community organizers was low; only one community organizer in New Orleans did not meet this commitment. A second organizer in Palm Beach County met her original commitment, but declined to stay for an additional year when the length of the program was extended. A third, hired as her replacement, performed poorly and was dismissed.

14. The hiring decisions were not controversial. In all cases, the local interview committees chose the candidates that Eichler and the local coordinators preferred from the pool seen by the committees. Nevertheless, the fact that a broad spectrum of locally respected people had backed the chosen candidates gave those candidates credibility and could have been used to provide the program with some "cover" if one of the hires had turned out to be a mistake—because they were "the community's choice." This is one of many examples in which well-thought-out actions were used to build a position of strength for the program; in this instance, the "insurance" was never needed.

15. This was especially important in New Orleans, where many families have lived in the city for generations, neighborhood identification and loyalty are strong, and outsiders are often viewed as "not understanding how things get done in New Orleans."

16. Barrera worked in Oakland with someone he later recommended to Eichler and who was eventually hired for the coordinator position in Baton Rouge.

17. As discussed elsewhere, the program plan was to provide technical assistance in real estate development through a combination of national and local technical advisers and LISC program staff.

18. Once the program was under way, training became an ongoing process that alternated between short, focused training sessions with Eichler and/or local coordinators and carefully supervised work in the field. This customized on-the-job approach suited the program well for several reasons. First, it was tailored to the diverse backgrounds and orientations of the staff and the different political and neighborhood dynamics in each site and targeted community. Second, it meshed with the nature of the skills staff needed to acquire—the basic concepts employed in consensus organizing were not complex; the complexity comes in applying them well in multifaceted and dynamic settings. Finally, it allowed formal training and coaching on specific topics to be provided at the time specific skills were needed, that is, when staff felt a need for them and when acquiring them built the organizers' confidence and motivation.

19. Some were also strategically important. For example, Pleasant City is *the* poor African American neighborhood in West Palm Beach, the largest city in its county. A decision not to work in what would appear to be the most obvious choice would have sent a negative signal about the program. Gert Town is home to Xavier University, a prominent local African American institution and a Seedco site. South Little Rock is the home of the program officer from one of the lead local sponsors, the Winthrop Rockefeller Foundation, and also of some prominent local African American officials in the city.

20. This process and its consequences are discussed in more detail in the next chapter.

21. Although the goal was to select six neighborhoods, the three organizers were not constrained to recommend two apiece. In Palm Beach County, for example, one organizer recommended all three of his neighborhoods, whereas another recommended only one. Although everyone recognized the advantage of maintaining continuity of staffing in each neighborhood, organizers were reassigned when necessary (after neighborhood selections were made) to achieve an equal distribution of work: two neighborhoods per organizer.

22. Although the existence of these groups demonstrated volunteer interest, it also complicated the issue of neighborhood selection because the development team had to decide at the same time whether to try to work with these groups rather than starting from scratch (the safer approach). Lake Worth's group had just been organized. The group in Pleasant City had a longer history but limited development experience. Neither group had broad-based community representation, which suggested potential benefits from development team assistance. The team's experience in working with these preexisting groups is discussed in the next chapter.

Organizing CDCs
and Developing
Indigenous Leadership

Objectives and Strategy

A primary objective of the demonstration program was to organize strong, effective CDCs in targeted neighborhoods. The program sought to build organizations that would develop technical and political capacities and would be under the control of community volunteers who were broadly representative of their communities and had a long-term stake in their neighborhoods. To that end, the development team identified in each neighborhood a diverse group of volunteers to form a CDC board and assisted them with their initial real estate project. As the board executed its project, the development team helped board members develop leadership and organizational capacity and acquire technical skills. Successful completion of the initial project was intended to lay the foundation for a continuing stream of community development activity by the CDC and to demonstrate the impact that the development of community leadership, organizational and technical skills, and working relationships with external sources of support could have.

The consensus organizing approach used by the development teams was designed to spawn CDCs led by diverse and cohesive groups of volunteers that were representative of their neighborhoods. The belief was that boards that mirrored the racial and ethnic composition of the community could give new CDCs credibility, both within their communities and with prospective external supporters. Representative boards were also likely to choose activities and projects that benefited the community as a whole.

Identifying individuals who would commit to making a significant investment of time and energy (sometimes referred to by program staff as *quality volunteers*) was critical. Often this meant facilitating the emergence of new leaders in targeted neighborhoods. Ideally, the new leaders would be relatively unencumbered by responsibilities to other organizations and would be willing to work together with other volunteers.

The organizing process was designed to encourage the sharing of responsibility among board members. It also sought to motivate volunteers both to accomplish project tasks and to continuously reach out to the community. This broad participation and outreach was intended to help boards attract committed volunteers, regenerate over time, and sustain their legitimacy. It also appealed to outside supporters who were drawn to the idea of investing in those who invested in themselves and their community.

Recruiting volunteers who understood and accepted the idea of starting the CDC's work with an initial real estate project was also critical. It was the program element intended to ensure that each CDC's first project was one that LISC was well positioned to support both technically and financially. The real estate focus provided each new board with a common goal around which to focus its energy. Real estate projects posed a real challenge for volunteers because they are complex, are technically demanding, and require effort sustained over a multiyear period.[1] Their successful completion was designed to build confidence among the volunteers, while winning them respect both in their communities and among external supporters. This combination was intended to lay the foundation for sustained relationships in support of an ongoing community development agenda that could (and presumably would) broaden beyond its initial housing focus.

Organizers used residents' desire to effect change in their neighborhoods, the pragmatism of the development team approach, and the availability of local LISC support to elicit volunteer commitments to join CDC boards. A major tactic used in recruiting volunteers was to emphasize the connection between the responsibility of CDC board membership and the rewards of that effort. The reward for volunteering and assuming responsibility was a chance to improve their neighborhood and to have a strong influence on the development activities undertaken. At the same time, volunteers would gain valuable experience and knowledge on a broad range of real estate, development, and organizational issues (i.e., they would enhance their technical capacity).

Once volunteers had been recruited and CDC boards had been organized, the focus of the organizing process shifted to four complex organizational issues. The development team had to effectively manage a series of "inherent tensions." They included:

1. Broadly based board participation *versus* centralized effort by a few knowledgeable and highly committed volunteer board members;
2. Providing volunteers with enough information and perspective to make sound policy decisions, *but* preventing them from becoming overwhelmed by the size and complexity of their task(s);
3. Preventing strong personalities or individuals with expertise from dominating the process, but doing so *without* the development team appearing to control the boards themselves;
4. Supporting non-real-estate projects or activities to fill any delays in the real estate production process *versus* diverting volunteer attention from the primary initial task (i.e., the real estate project) and slowing its progress.

Underlying each of these four tensions is perhaps the most fundamental challenge of all organizing efforts—for organizers to actively manage a process that shifts meaningful expertise, capacity, and control to groups *without* being overly directive and influential so as to dampen volunteer engagement.

Activities to Implement the Strategy

The process the development teams used to build broadly based, participatory organizations began with outreach that made deliberate efforts to bridge across existing neighborhood social networks. The community organizers assumed the frontline responsibility for crafting the CDC board, culling the talents of local volunteers, and molding a cohesive unit that would take on the responsibility of pursuing development projects. Organizers identified and sought to recruit representatives of six neighborhood networks: homeowners, renters, businesspeople, religious institutions, service organizations, and local institutions such as schools and hospitals.[2] They also placed great emphasis on creating boards that were representative along racial, ethnic, and gender lines.

In each neighborhood, the community organizers identified a core group of "founding" board members; typically this was the group that attended the press event announcing the target neighborhoods. Starting with this nucleus, the organizers worked to enlarge and diversify boards. This involved expanding the network of contacts the organizer established during the neighborhood assessment and using both that network and the board members already committed (i.e., the founding group) to identify other prospective candidates for boards. As the organizers received commitments from volunteers, the local coordinators monitored the composition of each developing board and, when necessary, pressed the organizers to "round out" their boards by identifying members of underrepresented groups.

Engaging a number of active volunteers and ensuring that they represented a range of community interests was essential both to ensure the organization's credibility and to develop a network of neighborhood leaders, not just a few individuals. It was also important because much of the CDC's work would be done by volunteer board members. If the board was too small, board members were at risk of "burning out," thus compromising the CDC's long-term viability.

The development teams segmented the organizing and development processes into clear, discrete steps with ambitious but reasonable deadlines. This staging was used to increase the board members' organizational and technical capacities without their feeling overwhelmed by "all the things they had to do." Successful completion of each step helped build the board's confidence and credibility. Development team staff provided support (including access to technical assistance) at each step. Most notably, staff structured their participation to allow (and sometimes to press) volunteers to assume responsibility for moving the process forward; this was critical to ensuring that the board developed ownership and control of the CDC. After the organizers identified a board nucleus, they helped guide volunteers as they recruited additional members, created bylaws, applied for nonprofit tax status, and elected officers. The newly formed group then held a town meeting, identified a target area and community development priorities, selected a real estate development project, and formed four project development committees (site and legal, marketing and counseling, design and construction, and finance).

The four project committees, assisted by the organizers and by members of the local technical team, undertook most of the work required to prepare the CDC's request for LISC predevelopment funds; the board as a whole provided policy direction and oversight. The committee structure allowed the CDCs to pursue multiple tasks simultaneously, keep them manageable, and share responsibilities among volunteers. By the time they requested predevelopment funding from the LISC Local Advisory Committee, the CDCs were expected to have completed a marketing study, established relationships with lending institutions, and identified properties to be acquired.

During this process, the development team tried to pair CDCs with individuals from a local technical team—professionals knowledgeable about real estate development and willing to work (typically at a reduced fee) with the largely inexperienced volunteers. After predevelopment funding was secured, the board selected and contracted with its own professional project team (e.g., architect, contractor), oversaw its work, began property acquisition, and obtained private and public-sector funding.

Throughout this process, the community organizer served both as staff support for the board and as a coach—helping volunteer board members learn to work effectively together within the framework of a nonprofit board and its committees, maintaining their motivation, and sustaining outreach to the broader support community. Thus, the organizer was the CDC's main link to the development team and larger support community.

The local coordinator provided day-to-day supervision of organizers, gave technical and tactical advice, and supported the organizers by attending key meetings, establishing contacts for the CDCs with prospective supporters, mediating conflicts, and, when needed, backing up the organizers. Michael Eichler helped to train the organizers, both initially and at each major step in the organizing and development process, and provided ongoing strategic advice and backup to the coordinators. For the most part, Eichler was a resource that the coordinators could use at their discretion, so his degree of active engagement varied across the sites. In Palm Beach County (the first site to start up), the local coordinator used Eichler extensively, whereas in Little Rock, the coordinator operated relatively independently. In New Orleans, the local coordinator used Eichler more than the local coordinator in Little Rock, but less than in Palm Beach.

Findings From the
Implementation Experience

The findings from the implementation experience fall into two main categories: early organizational outcomes and the dynamics that generated those outcomes.

Early Organizational Outcomes

The initial organizing component of the program worked well in all three sites. In each site, the development team organized broadly based CDCs with at least 10 members[3] and guided them as they applied for nonprofit tax status, hosted town meetings, elected officers, selected initial real estate projects, conducted marketing surveys, and met with lending institutions (see Table 5.1). Of the 19 neighborhoods targeted by the development team, all formally organized a CDC and all but one successfully petitioned the LISC Local Advisory Committee for predevelopment funding.

TABLE 5.1 Demonstration Program Timeline

							Duration of Program (in Quarters)										
	1[a]	2	3	4	5	6	7	8	9	10	11	12	13	14	15	16	17
DT Coordinator Hired	P, L, N																
Initial Fundraising Event[b]	P	N															
First LAC Meeting		P, N	L														
Neighborhood Selection Begins		L, N	P														
DT Announces Selected Neighborhoods			P, L				N										
First CDC Elects Officers			P, L														
First CDC Town Meeting			P	L, N													
First CDC Selects Project				P, L, N													
First CDC Approved for Predevelopment Funding					P, L, N												
LISC Program Director Hired	N				P			L									
First CDC Obtains Bank Financing					L		P					N					
First CDC Approved for Public Sector Funding					L		P	N									

Duration of Program (in Quarters)

	1[a]	2	3	4	5	6	7	8	9	10	11	12	13	14	15	16	17
First CDC Acquires Property						L		P			N						
Coalition Committee Formed							L			P, N							
LAC Extends DT								P, N									
First CDCs/ Groundbreaking								L		P			N				
Coalition Selects Officers										L	N	P					
First CDC Ribbon Cutting											L		P	N			
Coaliion Bylaws Adopted												N	P				
DT Coordinator Leaves									L				P, N				
Coalition Hires Executive Director										L				P			
Coalition Disbands													N				
Executive Director Leaves																P	L

NOTE: L= Little Rock; P = Palm Beach County; N= New Orleans.

[a]Quarter 1 begins when the DT Coordinator is hired: Palm Beach, May 1991; Little Rock, January 1992; New Orleans, April 1992.

[b]In Little Rock, the Initial Fund Raising Event occurred two quarters prior to the Hiring of the Coordinator.

The Dynamics of the Organizing Process

The key element of the organizing process appeared to be getting the newly forming CDCs off to a good start. Two components were essential to accomplishing this goal. The first was forming a board of adequate size to manage the work to be done. This was the program element most directly affected by the decision about whether or not to work with preexisting community groups. The second key element was ensuring that the group elected a strong cadre of officers—particularly an effective president.

As noted in the previous section, the community organizers did a reasonably good job of assembling initial boards of the size and composition desired. There were some exceptions, most notably (but not exclusively) in New Orleans; these exceptions were all in neighborhoods where the program sought to organize a new group from scratch. Problems of this type tended to surface early, for example, low activity and energy levels around the organization of the town meetings. In these cases, the organizer responsible for the neighborhood was motivated to increase his or her level of effort with the group and strengthen its membership and performance. The coordinator's response was typically to praise the organizers who had organized good groups and the volunteers who pulled their weight, and to continue to set tight deadlines for "next steps" that the "underperforming" groups were expected to achieve in an improved fashion. The combination of praise for work well done and "embarrassment" for underperformance, plus both the opportunity and the perceived need to "do better," runs throughout the program, and for the most part worked well during the interim organizing phase.

A few groups did not respond to this tactic, or responded but later relapsed. The issues involved seem idiosyncratic. In Palm Beach County, the high level of civic engagement in Del Ray Beach (which had initially made the community look like an attractive site) meant that many volunteers had competing civic commitments and could not give the new CDC the time it needed. Despite repeated efforts by both the organizer and core board members to expand the board, it remained small. In a New Orleans community, Gert Town, the organizer found many committed residents, but many were elderly and simply lacked the energy and optimism that a sustained development effort required.

■ Working With Preexisting Groups

The program's experience working with preexisting community groups was mixed. There were five cases in which the development team elected to work with preexisting groups (e.g., a homeowners organization or a church-based group). In each case, the team recognized that this decision entailed risks, because elements of the existing groups (e.g., their membership, operating style, or agenda) differed in important respects from those articulated in the consensus organizing model. Nonetheless, in those cases, working with existing groups seemed better than any alternative. The local coordinator (sometimes with Eichler) met early with each of these groups to explain the types of changes they would be expected to make if they were to be selected; these typically included restructuring and diversifying their boards, allowing the reconstituted board to select the group's leaders, and agreeing to adopt the community development strategy employed by the program. Each of the groups agreed, and the organizer's task was to help the organization live up to its promises.

Managing group process in these CDCs presented different issues than those encountered in CDCs created de novo. The team's mixed experience illustrates both the opportunities and the risks of operating the program with existing community groups. In two cases, the transition to a new CDC proceeded smoothly because volunteers shared objectives and values with the development team. In addition, the existing groups' leaders set a strong example by welcoming the development team, and potential problems arising from differences were addressed up front. In the three other existing groups, issues were not dealt with adequately early on, but eventually had to be addressed (often with negative effects on the progress of the groups' projects) or the groups became less functional over time.

The opportunities and risks of working with preexisting groups are clearly illustrated by Holy Cross in New Orleans and Pleasant City in Palm Beach County. Holy Cross CDC was organized from a core group affiliated with a neighborhood association that included a mix of black and white residents. Because they were already a diverse group and valued participation, they were very comfortable with the prospect of broadening the board. In addition, they had adopted an organizational style in which numerous people were active in the group's activities. Real estate development was a new activity for them and led to increased participation. However, the motivation for participation increasingly was self-interest, for example, being the first "in line" for housing the CDC was developing. The CDC has carried on with its projects with some missteps, especially when their original organizer resigned from the program staff and the Holy Cross CDC was less than fully satisfied with his replacement. Holy Cross has been about average in production terms among the New Orleans CDCs. Yet, this has been a disappointment to LISC program staff and volunteers, given the consistently large number of volunteers involved in the group.

The Pleasant City Community Revitalization Corporation grew out of an organization formed by members of five West Palm Beach churches and led by a well-known, highly regarded minister, Reverend Tyson. He agreed to take the risk of "giving up" his organization in exchange for the opportunity to develop a new organization—over which he would have less control, but which would have broader community support and access to the resources and expertise of

that followed a strong choice. His willingness to take a risk for the sake of the community and his grace in embracing the principle of inclusion set a tone of commitment that served the group well. Where the program failed to guide the selection of CDC presidents effectively, difficulties followed. For example, although the coordinator and organizer in North Little Rock foresaw problems resulting from the Argenta CDC's election of officers, the community organizer resisted intervention. The board consisted of 11 white members and five African American members. One presidential candidate was white and the other African American. Program staff feared that votes would be along racial lines and that African American members would feel alienated and leave the CDC. The local coordinator suggested that the organizer encourage the two candidates to nominate each other, but she did not wish to intervene.[7] The white candidate was eventually selected as president. Following the election, a committee was organized with only white members. That decision and subsequent board actions contributed to African Americans on the board feeling that their contributions were not highly valued, and gradually African American participation on the board declined.

Newly assembled volunteer boards were commonly inclined, at least initially, to elect as president either people who were already prominent or people who they believed had relevant technical expertise. Such individuals were sometimes good choices, particularly if they had jobs that gave them the flexibility to do some of the CDC's work during regular business hours. In other instances, the obvious candidates were not the group's best alternative, and the organizer's coaching was aimed at helping volunteers understand the strengths of other members of the group.

The quality of leadership in the CDCs suggests that the organizers' coaching was effective—but they were not always successful. For example, Limestone Creek CDC (in Palm Beach County), despite coaching to the contrary from the organizer, selected as its first president a prominent professional who was a member of the CDC's initial core group. He did not understand the development team process and resisted learning and adopting it. (For example, he refused to set an agenda for board meetings and was often not available by telephone.) His lack of understanding quickly caused other volunteers to lose patience, and they replaced him to positive effect. Other CDCs rebounded less quickly. The presidents of both New Vision (New Orleans) and Central Little Rock CDC were selected because of their notable accomplishments while serving on other boards. Although their other commitments prevented them from keeping up with CDC progress, their egos caused them to attempt to control decisions. New Vision was the last of the New Orleans CDCs to hold its town meeting and receive predevelopment funding, and indecisiveness caused it to miss important early opportunities to acquire property. Although Central Little Rock CDC's first project was relatively simple (relocating a donated house to a donated lot), this

activity came relatively late and the CDC almost missed an important deadline
to make the project happen.

■ Broad Participation
   and Shared Responsibility

The development team entered target neighborhoods with the objective of
promoting broad participation and shared responsibility. Fostering sustained
participation and resident commitment was both a value-driven and an instrumental program goal. Each site presents examples of CDCs that gained credibility and strength through broad board participation. Limestone Creek CDC in
Palm Beach County, highlighted previously, solved its early leadership problems
and attracted the attention of the chair of the County Commission by making an
effective presentation on behalf of their community at a Commission hearing.
The Commissioner, whose district included Limestone Creek, was impressed
both by the fact that the volunteers were technically well prepared and by the
cross section of residents that had mobilized. At her own initiative, the Commissioner began meeting with the group on a monthly basis to keep abreast of their
progress, anticipate when they might require County assistance, and give them
advice. Over time, however, the president of the CDC, a retired contractor who
was able to devote substantial amounts of time to the group's project activities,
began making progressively more of the CDC's decisions, and board membership and participation waned. In response, the Commissioner stopped attending
meetings. After the CDC had built a number of homes and began reconstituting
the board by recruiting the new home buyers, her support revived.

College Station CDC in Little Rock and Holy Cross CDC in New Orleans
were nearly forgotten neighborhoods prior to the LISC program. College Station,
a partially unincorporated neighborhood, had little political standing. Because
the neighborhood had a high demand for housing units, the CDC decided to
build a new housing subdivision. This project entailed the construction of basic
infrastructure and therefore required city cooperation. The initial response of the
city official responsible for the distribution of Community Development Block
Grant funds was, "why would anyone want to live in College Station?" His office
recommended that the CDC's request be denied. However, the volunteers invited
the mayor on a tour of the proposed project; their high level of engagement and
ability to show they represented the neighborhood during the tour impressed the
mayor, which contributed to his advocating the project before the city council,
and it received city support.

Prior to development team activity, New Orleans had established a policy of
targeting housing funds to specific neighborhoods. Holy Cross had not been
selected as a targeted neighborhood. However, within a year of organizing
efforts, the city selected the neighborhood as a demonstration site for its new

housing policy. The city awarded the CDC Neighborhood Housing Improvement Fund dollars, and later selected the CDC to administer an owner-occupied housing rehabilitation fund in the neighborhood. Soon after the CDC was organized, Tulane University Medical Center selected it as a partner in developing a community-based clinic, after an evaluation of many neighborhood organizations in the city. The favorable treatment the CDC has received from these two powerful New Orleans entities is a testament to its ability to claim it represents the community.

Conversely, failure to maintain active board participation can undermine the CDC's effectiveness. The New Orleans program presents one example. The first president of Algiers Crescent CDC (a white male) was knowledgeable about development, but not very sensitive to the delicacy of the racial balance that held the CDC together.[8] He decided that the organizational structure suggested by the community organizer (following the program model) was cumbersome and discarded it. Instead of assigning tasks to committees, he assigned them to individuals, regardless of their committee membership. African American volunteers felt they received insubstantial assignments and decreased their commitment. Although other CDC board members and the organizer eventually convinced the president to resign, membership continued to decline. White volunteers left the board with seemingly little provocation, as they either felt they carried a disproportionate amount of responsibility or were uncomfortable with the racial tension on the board. Project development slowed, despite the fact that the CDC operated in a good environment for property acquisition, and the board continued to struggle to attract new members.

The effects of inadequate resident participation on organizational capacity were even more dramatic for Mid City CDC, another New Orleans group. The community organizer assigned to Mid City was never able to attract Latino residents to the board, although they comprised a substantial population in the neighborhood. Early on, the CDC president pushed the group to begin development without first holding a town meeting. Several volunteers, whose input was not invited, left the board. The board seemed to stop functioning after it received predevelopment funding from LISC and selected its initial development sites. The president did not maintain contact with the program staff, and the board did not appear to pressure him to do so. As interest and membership numbers waned, new volunteers became harder to find. By mid-1996, the CDC had still not acquired any property (or engaged in any other activities) and support of the local LISC program director had waned.[9]

■ The Importance of Process

Program experience suggests that in addition to having the traditional leadership skills of motivating volunteers, coordinating information among committees, and resolving conflicts, leaders of the newly organized CDCs *must be especially sensitive to process.* CDCs associated with the LISC demonstration that had presidents who refused to sacrifice process to get output and who managed group process well were less likely to beome dependent on the president to get things done. These CDCs also were less likely to be set back when conflicts arose or when activity caused volunteers to lose interest. Three examples illustrate the advantages of "process-focused" leaders.

The College Station CDC president in Little Rock valued participation and believed in the development team process. He foresaw problems and ensured continued momentum. The group relied on the committee structure and maintained a large board with active and regular participation. The president was not considered a prominent citizen and had not demonstrated leadership capacity prior to the CDC's formation. The selection of College Station as a target neighborhood was considered risky by the development team. However, the CDC continually surprised and pleased program staff and supporters and is considered among the strongest of the Little Rock CDCs.

The presidents of the Downtown CDC in Little Rock and Pleasant City CRC in Palm Beach both acted to resolve early conflicts, with a respect for ensuring that procedures were fair. Their style established ongoing respect for inclusion, continual board building, and diversity. Leaders in these and other cases set a positive example of how participation should be structured.

■ Maintaining Momentum

A central challenge to the program goal of broad participation and shared decision making was the ever-present goal of maintaining momentum. The program had an ambitious timetable, and, as discussed in Chapter 6, funders can lose confidence in the program if it does not produce "hard" results in a timely manner. The internal logic of the organizing process requires momentum, as well. In the absence of progress, volunteers can lose interest or lose confidence that the program offers a real opportunity for effecting positive change in their communities.

One of the program's most powerful tools for maintaining momentum is its strategy of staging—segmenting the processes of incorporating a CDC and planning and developing a real estate project into relatively small, discrete steps. Each step presents opportunities to move the program forward at both staff and community levels and to recognize and publicly celebrate progress. In addition, at the staff level, each step provides an opportunity to train the organizers (and the coordinator, if necessary) in tactics for which they have an immediate need; this can help build their confidence and also provides them with an opportunity to succeed, thereby increasing both their confidence and their motivation.

Finally, at the community level, the benefits of staging are most numerous. Each step is small enough that volunteers new to development (i.e., most volunteers in the demonstration) can understand its role and importance in the overall process without being overwhelmed (and hence discouraged) by the complexities of the full development process and can master the skills needed to complete their part of the work. Discrete steps that can be conducted in parallel via a committee structure (1) provide opportunities for technical training in which the entire group participates—a shared experience in which each individual gains some understanding of what the others will be doing; (2) allow many volunteers to be actively involved in the development process simultaneously— gaining expertise and confidence while making a contribution; (3) give the group the satisfaction of seeing the project progress much more rapidly than if tasks were done serially; and (4) provide a series of motivating deadlines and opportunities to celebrate accomplishments.

The value of this sequenced process is illustrated most powerfully by the experience of the one CDC that did not benefit from it. Early in the program, Gert Town CDC in New Orleans had been among the most promising groups. It was the first CDC to elect officers and hold a town meeting. Prior to the meeting, the CDC had 19 board members and a general membership of 150, which the volunteers built through door-to-door canvassing. A local church organized a fund-raiser for the group, and another organization offered to donate 15 properties. However, after the town meeting, the project manager (acting on his own rather than to following program's training) constructed a complex and ambitious timeline for the group that listed more than 50 tasks for the full board and committees to accomplish. The lengthy schedule confused and overwhelmed the volunteers, who felt LISC's expectations were too high. They lost enthusiasm and eventually stopped meeting. They were the only group (of the 19 organized by the program) that did not reach the point of applying for predevelopment funding.

Despite the benefits of the sequenced process, however, maintaining participation by a diverse group of volunteers did require a commitment to sacrifice the short-term benefits of consolidated decision making in favor of building capacity and cohesion (bonds) among CDC board members. The perceived tension is real. Real estate projects are necessarily complex, and the groups face strong temptations to forgo participation for more rapid production. Some volunteers have financial, legal, or development experience; some have more time to devote than others; some have personal agendas they believe can be fulfilled through the CDC; and most see a real need in their community for housing development and have been convinced that a completed project will enhance access to additional resources. All of these create both pressure and opportunity for more rapid progress if broad participation can be dispensed with.

There were many examples, in all of the sites, in which individuals with development experience attempted to dominate projects or push the CDCs to move quickly. The development team responded by (1) using the staged process to manage the amount of detailed information volunteers had about upcoming activities and (2) emphasizing that the development team process was designed to help the CDCs attract needed resources. This strategy was effective in some CDCs, such as McClellen (North Little Rock), but not in others, such as Argenta (in North Little Rock), where the CDC was well positioned politically and felt no need to demonstrate "representativeness" as a way to ensure access to resources.

The sequenced process and the committee system served most groups well until they had prepared a project proposal and received predevelopment funding from LISC. Up to that point, there were clear roles for many volunteers, and many were needed because in the absence of funding, they had to do most of the work themselves. They could do the work, following a demanding but not unreasonable timetable, if provided with adequate technical help. This help was abundant in Little Rock, and projects went forward at a respectable pace (although somewhat more slowly than the program's original, ambitious two-year timetable called for).

Unfortunately, the attempt to use local technical consultants to assist CDCs with their projects worked well only in Little Rock. The local coordinators in Palm Beach County and New Orleans had great difficulty finding and retaining a cadre of development professionals (local lawyers, architects, contractors, and planners who work with the CDCs for less than their usual fees) able to work effectively with the CDCs and willing to do so over a sustained period. For example, only one member of the nine-member technical team originally recruited by the local coordinator in Palm Beach County stayed in that role over the initial two-year demonstration period, and replacements were not identified.[10] Similar, but less acute problems arose in New Orleans; having seen the problems that had arisen in Palm Beach County, the New Orleans coordinator was quicker (but not quick enough) to retain additional technical help on a consulting basis. Even in Little Rock, where the coordinator had success in recruiting and retaining technical help, the number of individuals involved throughout the two-year start-up period was still small.

The absence of adequate technical support had two main ill-effects. Predictably, project progress slowed, tiring and discouraging many volunteers. Less obviously, the organizers (and to a lesser extent the coordinators) were increasingly pressed to provide help with technical tasks for which they lacked adequate training. They were, after all, selected for their skills and commitment as organizers, not developers, and only one had any directly relevant training. However, they were committed enough to the program and their volunteers to do whatever they could to be helpful. The results were that their organizing activities increasingly got short shrift, and the organizers were vulnerable to losing the credibility and rapport they had enjoyed with the groups when they were unable to provide the technical help the volunteers needed.

The tension between process and product was felt most acutely after the groups received predevelopment funding. In several CDCs, a subgroup of volunteers, typically two to five members, assumed responsibility for project decisions, and the committee structure disintegrated. Such a core group emerged in the CDCs that were most productive: Boynton Beach and Limestone Creek CDCs in Palm Beach; McClellan, Argenta, and Downtown CDCs in Little Rock; and Bacatown CDC in New Orleans. For example, prior to receiving predevelopment funding, McClellan CDC's president had worked to ensure that all members participated, particularly ones like herself who had not had past leadership experience. She insisted that no officers chair committees and that progress be paced so that everyone understood developments. However, after the CDC's complex proposal for a multifamily development was approved by the LISC Local Advisory Committee, she attempted to control the process, demanding that she approve each committee's work. Eventually, three of the four committees ceased to function, and the finance committee made most of the decisions.

The leadership of the Downtown Little Rock CDC also made a conscious decision to control project decisions. Only three or four CDC members were involved with project details, and progress was dependent on the president and attorney. They felt the multiunit project was too complex for others to understand and the committee structure too difficult to coordinate. Their attorney commented, "at this point, you don't really need a lot of volunteers; amateurs get in the way." Nevertheless, between 12 and 15 members routinely attended board meetings and felt engaged because the president kept them abreast of general developments, and members believed they could rely on him to make sound decisions.[11] In contrast to the Downtown Little Rock CDC, Bacatown (in New Orleans) CDC's membership declined while the group's project progressed. Five members were involved in project details, and the CDC's president did not work to keep other volunteers informed. When Bacatown entered a joint venture with a private developer, volunteers "disengaged" from development activities as the developer pushed projects forward on his own with little contact with volunteers.

This pattern raises difficult issues about real estate as a point of entry for programs designed to increase resident commitment, capacity, and control. The benefits in terms of earned respect from external supporters (and the community itself) when results are produced are reasonably clear. On the other hand, the complexity and technical difficulty of the development process, combined with the length of time over which volunteer energy must be sustained and the clear difficulty of engaging CDC board members meaningfully throughout the process, is equally apparent. The examples of CDCs like College Station and Downtown Little Rock illustrate that solid progress on the full range of goals is possible, but the challenges of managing the process well are formidable. The presence of adequate and timely technical support is essential—and its local availability cannot be counted on in localities where community development is relatively new.

Lessons

The program strategy of segmenting the organizing and development process into discrete steps proved central to its initial success. It provided a critical tool not only for keeping a complex set of tasks organized and moving forward, but also for managing the organizing portion of the program. The organizers used it to keep forceful individuals from dominating the boards, to motivate volunteers, and to build volunteer confidence and skill. This strategy was easiest to implement in the early stages of the program when many tasks needed to be performed simultaneously (making it possible to spread the work broadly and keep many people involved), but became more difficult when the CDCs entered the construction phase of development.

A key lesson is that successful organizing and sustained volunteer participation depend on a program's ability to deliver both ongoing organizing help and adequate technical support to allow the volunteers to do their project work well and on time. The complexity of the real estate development process, however, often made volunteers heavily reliant on technical assistance and tended to undervalue the importance of ongoing organizing support.

The experience of working with preexisting organizations demonstrates the need to address and resolve early on any concerns or potential mismatches. This seems especially important when one recalls the fact that the groups with which the program decided to work had been carefully screened and had explicitly agreed to make changes that would move them within the program's operating and philosophical framework.

There is a clear need for ongoing organizing, to achieve the difficult goals of enhancing resident commitment, capacity, and control. Even after boards have acquired a solid level of real estate skill, ongoing organizing is necessary if the boards are to remain broadly grounded in the community. In order for boards to maintain membership, attract new members, and survive challenges, volunteers must be reminded of the importance of broadly based participation and adhere to established procedures and values.

CDC leadership, especially the board president, plays a key role in maintaining broad participation and board diversity as well as in ensuring production. Despite the perceived tension involved, staff must carefully guide the officer selection process (especially early in the CDC's life) and sensitize leaders to the benefits of participatory processes and following agreed-on procedures. This is a delicate job, but it clearly can be done—and not doing it effectively appears to carry significant costs.

Notes

1. In the abstract, one could imagine other start-up activities that would have the attraction of being less intense than real estate development and more easily executed by volunteers new to community development. However, real estate development, particularly housing, is LISC's "bread and butter" expertise—a fundamental building block of the national track record that made them such an attractive sponsor for this demonstration. Embarking on a different program activity (e.g., youth employment or drug treatment) that did not build on their organizational strengths would have increased program risk.

2. Representatives from the latter four networks do not necessarily live in the neighborhood, but have a significant stake in community development efforts. Residents predominate on the boards of all the CDCs in the program.

3. The organizers' goal was to develop boards that each had between 12 and 18 members, and they fell short of this target in only two neighborhoods. The median board size at the time the CDCs formed was 16.

4. About one third of the CDCs had engaged in non-real-estate activities that involved other members of the community, for example, neighborhood cleanup or youth programs in collaboration with other organizations, and fund-raising events such as bake sales or car washes.

5. The overwhelming majority of the people of color in the three demonstration sites were African American. The only exceptions among the targeted neighborhoods were Lake Worth in Palm Beach County, which contains a small but growing Haitian community, and Mid City in New Orleans, which has a sizeable Hispanic population. These subpopulations were not well represented on the CDC boards, and the volunteer pool included only two Hispanics, one of whom was the founding president of Lake Worth CDC.

6. They saw, for example, vacant storefronts on the main commercial streets, rental housing that was showing signs of deterioration, and the influx of a poor, Haitian immigrant population.

7. This illustrates the cost that is sometimes paid when the program management strategy is to assign both authority and responsibility, holding staff accountable for results. The coordinator felt the organizer's judgment was mistaken, but allowed her to make her own mistakes and then be responsible for addressing any problems that arose.

8. Algiers Crescent (a New Orleans neighborhood) is sharply divided between a mainly white population in the gentrified part of the community and a predominantly black population in the balance of the neighborhood. Although both groups identified strongly with the neighborhood, the two segments of the community had never worked on anything together and tended to view one another with mutual suspicion and mistrust.

9. Subsequent to our final round of data collection, the Mid City CDC revived and resumed work on its project, thereby regaining LISC's willingness to work with them.

10. In Palm Beach County, Lynda Harris (a partner in a prominent law firm) helped the new CDCs with their articles of incorporation and other legal matters. Three other members of the original technical assistance team in Palm Beach County changed roles; one was hired by one of the CDCs as its project architect, one was hired as the local LISC program officer, and one became the president of one of the CDC boards.

11. Early in the CDC's history, the president had faced several conflicts and managed to keep members from becoming alienated, and his diligence set an example for all board members. The project also received community support because the CDC worked to keep the community informed about its development. In addition to holding annual town meetings, the CDC updated the community and elicited general input after it received predevelopment funding and again just before the project was complete.

CHAPTER 6

Building Relationships With the Private Sector

Objectives and Strategy

The LISC demonstration had both short- and long-term objectives in building relationships with the private sector. The primary short-term objectives were to cultivate commitment to and support for the LISC demonstration approach to strengthening low-income neighborhoods, enlighten the support community about conditions in targeted neighborhoods, and initiate a series of positive contacts between the private sector and the residents of those neighborhoods. The long-term objective was to create bridges between the private sector support community and low-income communities and thereby foster a community development support system with significant private sector participation. This entailed increasing the private sector's financial, technical, and political support for community development activities and enhancing the ability of neighborhood residents to act effectively with the private sector in activities that would benefit their communities.

The long-term objectives were very similar to LISC's objectives in all the metropolitan areas in which it is active. What made the demonstration unique for LISC was the timing of involvement and the range of private sector involvement requested. In "traditional" sites, LISC enters after the foundation for a community development program is already in place. This was not the case in the demonstration sites, where the private sector—particularly lead program sponsors and hosts—was asked to contribute to the program in multiple ways, including contributing to the building of bonds and bridges that would provide the foundation with ongoing community development efforts.

The demonstration's approach to building relationships with the private sector is most distinctive for its emphasis on consensus, that is, identifying opportunities for mutual benefits. Other community organizing approaches tend to focus less on the private sector than on gaining public sector support. The community organizing efforts that do focus on the private sector tend to take a more confrontational approach, such as Community Reinvestment Act (CRA) challenges.

Initially, cultivating solid relationships between project staff and the private sector was the primary focus. Later, the development team staff gave priority to fostering direct relationships between the private sector and the residents of targeted neighborhoods. Early in the process, program staff used the national reputation of LISC and the support of prominent local sponsors to build ties to the leading private sector organizations in each site. These ties were then used to (1) build financial support for LISC's core program and for the development team; (2) increase local private sector leaders' understanding of community development issues and the demonstration's approach; and (3) gain commitments of assistance from private sector participants, with a priority given to activities that required some contact with development team staff and eventually, for some participants, direct contact with neighborhood residents.

Activities to Implement the Strategy

The demonstration's primary mechanism for forging bridges with the private sector was to identify a variety of roles in which members of private sector organizations could participate in the program. Program staff sought to cultivate both formal and informal relationships and to strengthen those relationships (and hence commitment to the program and community development) through frequent interaction. Initially, the relationships were between private sector participants and LISC core program and development team staff. Later, after private sector participants had become more familiar with community development, developing direct relationships between neighborhood volunteers and the private sector increased in importance.

Formal program roles that private sector organizations and individuals assumed included (1) lead sponsor in bringing the LISC demonstration to the site, (2) host organization that provided office space and a well-regarded local "home" for the development team, (3) contributing to LISC and serving as a member of the Local Advisory Committee, (4) serving on the committees that interviewed candidates for the local coordinator and community organizer positions, (5) being a member of the local coordinator's strategy committee, and (6) providing private financing for CDC development projects.[1]

Informal roles included suggesting candidates for the community organizer positions and for the local technical team; being available for ad hoc consultation with program staff; interaction with CDC board members about development projects; working with development team staff and neighborhood residents to extend the support for development team efforts in the private, nonprofit, and public sectors; and advocating and supporting community development efforts in the public arena, including local media.

The process of building private sector bridges began during the site assessments, which included numerous interviews with representatives of leading private sector organizations. In their interviews, the LISC assessment team (Michael Eichler and Richard Manson) attempted to make explicit their expectations of the private sector. The first was that the private sector would have to raise funds to help support the local program. Introductions and support from prominent local sponsors were central to creating a positive fund-raising climate. Importantly, the interviews highlighted additional contributions required from the private sector. These included technical help for the volunteers and use of their political influence to garner additional resources for community development efforts. Finally, the interviews also highlighted that the private sector would have the opportunity to ensure high program quality and accountability through membership on the LISC Local Advisory Committee and contacts with LISC and development team staff.

The site assessment team made it clear to potential private sector supporters that their organizations could benefit from participating in the program, because this effort was not intended to be just charity. For example, banks could benefit from expanded demand for loans and fulfillment of CRA obligations, and other businesses could benefit more generally from improved conditions and a higher quality of life in their metropolitan areas. Contributors could also benefit from enhancing their reputation as good corporate citizens.

Once it was clear that the funding target would be met and a local coordinator had been hired, program staff and the local sponsors jointly selected a program host. The host's formal role was to provide office space and support for program staff, but a key criterion in the choice of a host was the organization's credibility with the private sector, and hence its ability to provide access to other private sector organizations that could support the program's objectives.

As in other LISC sites, the LISC Local Advisory Committee was the primary formal mechanism for sustaining private sector participation in the demonstration sites. Each of the institutions that contributed at least a specific amount (typically $10,000) to the local LISC funding pool had a representative on the Local Advisory Committee. These committees, which meet quarterly, oversee local LISC activities and make decisions about grants and loans to CDCs.

Regular meetings and a substantive community development agenda make the Local Advisory Committee a prime vehicle for strengthening ties to the private sector. The meetings provided an opportunity for members to learn about the activities and progress of the development team, about other LISC activities (both locally and nationally), about specific community development issues and activities in the targeted neighborhoods, and about the community development process more generally.

Eichler engaged a few private sector participants directly in early program activities by recruiting them to be on the committees that interviewed candidates for the position of local coordinator at each site. He selected individuals based on his assessment of their understanding of the program and of the weight their views would carry with their private sector peers. Local coordinators used a similar approach when they assembled committees to interview candidates for the community organizer positions.

Coordinators were supposed to organize strategy committees to advise them on overall program strategy and assist them with a variety of program implementation issues. These committees were to include members of the private-sector support community—from both the LISC Local Advisory Committee and others—and help coordinators in building and enhancing local community development infrastructure and support. Strategy committee members were asked to advise and work with CDC volunteers on activities organized by the coordinators, primarily activities related to their real estate projects.

The development team attempted to use an ongoing series of active contacts to promote enduring relationships between neighborhood residents and private sector organizations. The rationale for this was that repeated positive contacts would help break down barriers, help change detrimental misconceptions and behavior both in the support community (toward the residents of low-income neighborhoods) and among low-income residents (regarding the potential of their relations with the private sector), encourage cooperative effort, and help facilitate identification of opportunities for mutual gains.[2]

Most of the program-facilitated contact between neighborhood residents and the private sector was centered on the CDCs' real estate project activities. The underlying rationale was that successfully completed projects would be acknowledged and valued not only by residents of the target neighborhoods, but also by members of the private sector who understood how difficult such projects are and respected the skills needed to accomplish them.

A core element of the demonstration's strategy in building bridges with the private sector was to "mainstream" community development by facilitating "regular" working relationships between neighborhood residents and private sector organizations and individuals. Real estate development seemed well suited to this approach. The process of planning and completing these projects required extensive interaction over a considerable period, and thus provided opportunities to overcome private sector misconceptions about the residents of low-income communities. This was especially true in the financial sector. An objective with banks that provided financing for CDC real estate projects was to establish "day-to-day" working relationships between neighborhood residents and professional staff at local banks—the same types of mutually respectful and beneficial relationships enjoyed by other developers. The CDCs' initial development projects were thus both a means and an end.

Findings From the Implementation Experience

During the initial stages of implementation, program staff made significant progress in building relationships with the private sector. This was accomplished primarily through site assessment and fund-raising efforts, cultivation of good relations with the local sponsors, and the establishment of the LISC Local Advisory Committees. Other relationship building strategies—including the use of strategy committees and the use of private sector representatives to interview development team candidates—were generally less intensive and less successful. Over time, particularly as real estate projects lagged behind schedule in two sites, relationships with the private sector fell below expectations.

Program Start-Up Laid a Strong Foundation

The site selection process laid a strong foundation for building bridges with the private sector. During these assessments, Eichler and Manson met with representatives of most of the leading corporations, financial institutions, and foundations in the metropolitan area. They had ready access and immediate credibility based on the efforts of the lead sponsors to bring LISC to the sites and based on LISC's national reputation. LISC had particularly strong credibility in the private sector because of the business leaders on its national board, because it distributed funds to community groups using standards similar to those employed by banks (e.g., financial soundness, strong marketing plans), and because of its excellent financial track record in funding CDC-sponsored projects across the country. In addition, several private sector leaders in Palm Beach County and Little Rock had had positive experience with LISC in New York City or other places before moving to one of the demonstration sites.

Eichler and Manson impressed their interviewees with their commitment, knowledge, the pragmatism of the community development approach they were advocating, and their frankness and personal charisma. For example, Len Lindahl, at the time chair of the Palm Beach County Economic Council and later

TABLE 6.1 Initial Fundraising in LISC Demonstration Sites

Site	Total Raised Locally	Number of Initial Contributors
Palm Beach County	$560,500	20
Little Rock	$740,382	17
New Orleans	$1,020,000	13

city had a national reputation for poorly managing its community development efforts, including funding "do-nothing" community groups and using city funds for political patronage. In addition, Entergy was worried about housing conditions and their customer base. Freeport McMoRan was concerned about their inability to recruit high-skilled professionals and other major corporations to the city. No major new employer had moved to the city in nearly 20 years, and one of the most often cited reasons was the decline of neighborhood conditions just outside the central business district.

Local program hosts—the Palm Beach County Economic Council, the Little Rock Chamber of Commerce, and the Greater New Orleans Foundation—provided respected local "homes" for the program. This included office space in a location familiar to program supporters, modest office support (e.g., receptionist, message service, access to photocopy and fax machines) until the program generated enough activity to warrant its own support staff and equipment, and people on whom the coordinator could call for good information about the political, economic, and social context.

The Economic Council in Palm Beach County and its director, Dale Smith, provided a substantially greater degree of assistance for development team efforts than the hosts in the other two sites.[7] The fund-raiser for the Economic Council invested considerable time providing logistical support for the fund-raising effort. Smith himself developed a deep interest in the program, and acted as a resource for the local coordinator, helping her strategize and supporting the development team's efforts in a variety of informal ways. For example, he provided access to technical expertise needed by the Limestone Creek CDC when it opposed expansion of a highway right-of-way in its neighborhood, and directly supported the efforts of the development team and the CDCs to build working relationships with county officials and staff.

The Role of the
LISC Local Advisory Committees

The LISC demonstration clearly helped build bridges between the private sector and low-income neighborhoods and broadened private sector support for community development in each of the sites. In Palm Beach, more than 20 representatives of private sector supporters served on the LISC Local Advisory Committee; in Little Rock, a dozen; and in New Orleans, nine. The visible service of such prominent members of the business community enhanced the program's credibility. Although the level of engagement varied across the sites—strongest in Little Rock, weakest in New Orleans—Local Advisory Committee chairs and numerous other committee members were active in supporting development team work. They promoted LISC's efforts within their network of contacts. They engaged directly with resident volunteers in the targeted neighborhoods. At strategic points in the program, they helped win public sector support for the program, engender positive press, and build further support from the private sector. In both Little Rock and Palm Beach County, individual members of the private sector emerged as strong and influential community development supporters.

In Palm Beach County, the largest contributor—the MacArthur Foundation—was strongly supportive of private sector leadership of the LISC Local Advisory Committee. The local program officer did not want the program to be seen as a creation of the foundation. Instead, MacArthur wanted to foster private sector ownership of, and responsibility for, the program and community development generally in the county. Hence, all three committee chairs (first Len Lindahl, followed by Bill Peterson, president of a local television station, and later Pat Cooper, President of Chemical Bank and Trust Company of Florida) have been corporate leaders.

In Little Rock, the largest contributors and most active fund-raisers—Worthen Bank and First Commercial Bank—preferred to play a less visible and active role on the advisory board and supported as the first chair the widely admired African American executive director of the Winthrop Rockefeller Foundation, Mahlon Martin.[8] They also supported the inclusion on the LISC Advisory Committee of a range of civic leaders to help build broadly based support for program efforts—including the mayors of Little Rock and North Little Rock, the Pulaski County Judge, and an official of the local NAACP. The two lead banks nevertheless remained actively committed to the program and highly influential.

In New Orleans, the first Local Advisory Committee chair was James Cain, Vice Chairman of the Entergy Corporation. Cain could not commit the time necessary as chair. He was succeeded by Louis Freeman, the head of the local RosaMary Foundation (the fourth largest contributor to the local LISC pool). Over time, the local LISC program director convinced a senior vice president of Freeport-McMoRan, Chip Goodyear, to assume the chair because he saw the need to increase private sector commitment and involvement in the program. This was not unanimously supported, because some on the Local Advisory Committee were in favor of continuing with a chair from the philanthropic community.

LISC Local Advisory Committee members have been an informal resource for the development team in numerous ways. Examples of individual board members assisting development team efforts include the work of the chairs of the Local Advisory Committees. For example, Len Lindahl in Palm Beach County provided much support for the development team through the Economic Council, which served as the program sponsor. His successor as chair, Bill Peterson (president and general manager of a local television station), helped LISC-organized CDCs receive positive press on the station. In New Orleans, the current chair's (Chip Goodyear) position as senior vice president at one of the city's most prominent companies gives high credibility to efforts. In addition, his visits to neighborhoods for groundbreakings add visibility and credibility to the program, as the media cover the events (and often highlight his role and support). In Little Rock, the first chair of the LISC Local Advisory Committee (the president of the locally based Winthrop Rockefeller Foundation) was successful in helping College Station in its negotiations with the mayor to secure infrastructure resources necessary for a LISC-funded development project.

Nevertheless, some opportunities to strengthen ties with the LISC Local Advisory Committees were lost. It appears that even though the relations with the private sector went beyond the standard LISC site, the potential of the Local Advisory Committees to foster deepening relationships with the private sector was not fully realized in any of the sites. There were different reasons for this in the different sites. However, there did appear to be some commonalities. The most progress was made through the LISC Local Advisory Committees in establishing program credibility and providing a forum for informed conversations (at least among key members of the support community). The area where the potential was least tapped was with representatives from the private sector assuming leadership on community development in general and program accountability in particular. This could have been accomplished with more constructive critique of staff and the newly formed CDCs (and later the coalitions in Little Rock and Palm Beach County). (See Chapters 5, 7, and 8 for more detailed discussion.)

Connections between LISC Local Advisory Committee members and local coordinators, particularly in New Orleans, were not as well developed as intended in the program design. This limited the opportunity for constructive critiques, as well as more general bridge building. The local coordinators in New Orleans and Palm Beach County never had quite the same credibility, standing, or influence with corporate supporters as Eichler had in all three sites or the local coordinator had in Little Rock. At least part of the later loss of momentum in all three sites could be attributed to the limited depth of engagement of private sector supporters.

Strategy Committees

A significant implementation weakness occurred with the strategy committees. The strategy committees were envisioned in the program design as providing vehicles for key support-community actors to interact with each other and with the coordinator and engage in programmatic effort on a regular basis. However, they were not effectively instituted in any of the sites (i.e., to strengthen bonds among supporters and build bridges between supporters and neighborhood volunteers).

In Palm Beach County, the strategy committee was the most active—meeting on a fairly regular basis during the program's first two years and with declining regularity thereafter. For example, members visited CDC boards before their first formal funding requests were reviewed by the LISC Local Advisory Committee. This provided the CDC board members an opportunity for personal contact with private sector leaders,[9] and, as importantly, gave strategy committee members a chance to be impressed by the development expertise and commitment of neighborhood residents. At the LISC Local Advisory Committee meetings that formally reviewed initial funding proposals, the strategy committee members were often the strongest advocates in support of funding. Strategy committee members were able to convey effectively (to the full Local Advisory Committee) the competencies and enthusiasm of the CDCs making the funding requests and the quality of the specific funding proposals. The personal contact that occurred during strategy committee visits to neighborhoods also resulted in unsolicited commitments. For example, during one neighborhood meeting, a bank executive committed to assign a staff person to work with a CDC on securing private financing and having that person attend CDC board meetings. Finally, near the end of the initial two-year program period, strategy committee members (along with other members of the Local Advisory Committee) were instrumental in gaining support and funding for an 18-month extension of development team efforts.

The strategy committee in New Orleans was less structured than in Palm Beach County and met only a few times. The Little Rock strategy committee was transformed into informal, yet regular, contacts between private sector leaders and the local coordinator.

Only in Little Rock did the coordinator do a consistently good job of orchestrating the ongoing involvement of private sector sponsors and supporters, because he kept key private sector agents informed and engaged in program efforts. He consistently reached out and met with key program sponsors and this paid off. In contrast, in both New Orleans and Palm Beach County, after the national program coordinator, Mike Eichler, reduced his role, private sector supporters seemed to be less informed and engaged.

Strategy committees could have been used more effectively as a way to connect the support community to the CDCs and educate the private sector about community development. The "lost opportunity" with strategy committees is reflected in the comments of two major program participants in New Orleans—the LISC program director, Rob Fossie, and a member of the LISC advisory board and lead program fund-raiser, Diana Lewis. Both spoke of the value of the strategy committee in New Orleans when it was employed at the initial stages of efforts. Both indicated that the Local Advisory Committee members who visited the development sites became the program's strongest supporters.

In all three sites, it appeared that the hiring of the local LISC program director contributed to a decline in substantive relationship building with the private sector. When the LISC local program directors were put in place, the LISC Advisory Committees became in effect "their committees," and this closed out some opportunities (built into the demonstration program design) for the local coordinator (and hence community organizers and volunteers) to have good access to Local Advisory Committee members. LISC hired program directors when local CDCs appeared almost ready to start generating project activity, so the increased focus on real estate projects was not unanticipated. What was not adequately anticipated was the diminished ability of local coordinators to keep issues of organization and leadership development on the Local Advisory Committees' agendas.

This dynamic was most prominent in Little Rock and New Orleans. In New Orleans, the LISC program director was hired at the start-up of program effort. This was done because it was determined that the site would be the most difficult to work in, and there was some hope (as suggested by LISC's planning documents) of getting some money "out-the-door" quickly to existing groups. In addition, the site had the largest up-front dollar commitment, so having another staff person on site was thought to be beneficial and affordable. In New Orleans, with the local program director in place early, the local coordinator had the greatest difficulty establishing an independent relationship with the private sector support community and using that to build bridges between low-income communities and the private sector in the city.

In contrast, in Little Rock, where a local program director was not hired until after the end of the two-year demonstration period, there was deep and broad engagement of the private sector skillfully facilitated by the local coordinator, Richard Barrera. After the local coordinator departed, the participation of LISC Local Advisory Committee members declined significantly because the LISC program director did not reach out and engage committee members in the same way. Some Advisory Committee members were dissatisfied with the new focus in Little Rock on production and declining leadership and organizational development that ensued on the transition from the local coordinator to LISC program director. In addition, some committee members felt that they were inadequately involved in decision making and that the program was losing accountability to local funders.

In Palm Beach County, the local LISC program director was hired earlier in the program than in Little Rock and later than in New Orleans, and this had mixed results. Although the local coordinator was able to retain positive relationships with private sector members of the Advisory Committee, her relationship and subsequently her credibility with committee members declined as the local program director took over some of the leadership of program effort.

None of the local program directors lived up to initial expectations regarding the short-term demonstration program objective of active engagement of the private sector in community development, that is, broad-based participation with ongoing learning and bridge building.

Sustaining Private Sector Support:
The Role of Real Estate Development

Interest in the program by local private sponsors was most intense at start-up and at key events and decision points. Sustained, broadly based involvement was hard to achieve. Some of the support initially provided by the MacArthur Foundation in Palm Beach County was effectively picked up by the development team host, the Economic Council. However, when Dale Smith left the council, that support—most notably guidance on strategy—faded. In Little Rock, one of the lead sponsors, the CEO of Worthen Bank, was distracted from his efforts with the LISC demonstration when his bank was acquired by an out-of-state institution. In New Orleans, the Greater New Orleans Foundation's involvement was reduced when the development team staff relocated to another office shared with the LISC program director and as the foundation started to focus its attention on other community development initiatives (e.g., the Ford Foundation Collaborative).

Continued active and visible support by private sponsors would have been helpful, particularly through the difficult stage of transition in all three sites (see Chapter 7). This was most evident during the formation of the coalition and the hiring of an executive director in Palm Beach County, but was also a factor in Little Rock, as the lack of explicit direction and guidance from the lead sponsor and other private sector supporters appeared to ultimately be detrimental.

There were significant setbacks in relationship building with the private sector as CDCs lost their initial momentum after the start-up of their real estate projects. The main problem in New Orleans and Palm Beach County was that real estate projects did not meet expectations—taking much longer than expected. This, as highlighted previously, was closely related to the problems in providing adequate technical assistance. Delays in real estate projects resulted

In contrast to the local coordinator, the executive director of the Little Rock coalition never aggressively reached out to private sponsors and was unable to gain their active support (see discussion in Chapter 7). He provided fewer opportunities for private sector supporters to effectively influence efforts. Instead, he seemed to take the position that it was not his role to reach out to private sector supporters, that if the private sector wanted to help they would come to him. The executive director was an African American from a lower-working-class background who had received his college degree from Arkansas Technical University in Russellville. He appeared to have difficulty feeling comfortable asking corporate leaders for assistance, even though some of the corporate leaders (including the first chair of the LISC Advisory Committee) were African Americans—suggesting that cultural issues were at least as great a factor as race.

Private sector bridging in New Orleans has been the least substantive. The reasons for this relate closely to the local context. The private sector in the city had less prior experience with community development than their counterparts in the other sites. This lack of experience appeared to contribute to unrealistic expectations about progress on real estate and the least recognition (of any of the sites) given to the accomplishment of the initial organizing of CDCs.

In contrast to Little Rock, private sector relationship building in New Orleans did not progress significantly after the initial financial commitment. The local sponsor and host, the Greater New Orleans Foundation, did not have the same standing or depth of connection with the business community as the sponsors in the two other sites. The main local fund-raiser used personal contacts to raise a lot of money from a small number of private sector supporters. However, both the depth and breadth of support were relatively modest—a few supporters contributed a lot of money.

The lack of breath and depth of corporate sponsorship in New Orleans seemed to limit bridge-building potential. The character of support proved to be detrimental to program development, because there was relatively little engagement by the small number of prominent funders and smaller funders were relatively insignificant.

As real estate development in New Orleans lagged behind schedule, instead of lead sponsors and other private sector supporters assuming greater responsibility and working with volunteers and program staff more closely, there was a tendency to point fingers at others—for example, national LISC, the local coordinator, the city—and to pull back from program effort. As disappointment with real estate project activity increased (after three years no group had broken ground), private sector support and confidence in program effort and staff suffered. It could be argued that relatively passive support for community development in New Orleans was systemic; however, part of what was promised by the LISC demonstration to private sector program supporters was systemic change (i.e., change the way community development was undertaken in the city).

Over time, a small number of LISC Local Advisory Committee members appeared to increase their comprehension of community development and of what was required in New Orleans to make significant progress toward achieving local objectives. However, this was accomplished at some risk of oversimplifying with a focus on real estate production. Private sector supporters of LISC in New Orleans (at the three-year transition point) challenged LISC to deliver in terms of real estate before they would consider contributing more, with relatively minor consideration given to organizational issues. After informed discussion and in light of the fact that for the most part the groups had made little progress on real estate, the local LISC program director in New Orleans hired two technical staff persons to help work directly with fledgling CDCs. The decision to hire staff was made without strong support from a broad base of sponsors.

In contrast to New Orleans, private sector supporters in Palm Beach County seemed to be more patient and interested in intermediate outcomes, as suggested by enduring commitment (even after problems with the coalition; see Chapter 7). The contrasting experience in Palm Beach County can at least in part be explained by some program supporters having had previous community development experience (and therefore more realistic expectations), as well as the patient and experienced sponsorship of the MacArthur Foundation.

Private sector support in Palm Beach County receded with setbacks in real estate development and problems at transition and with the coalition (see discussion in Chapter 7). However, in spite of these difficulties, major supporters have sustained their commitment to LISC efforts in the county. This indicates that there was some significant success in relationship building with the private sector in Palm Beach County. This is true in the other two sites as well. The second round of local LISC fund-raising generated firm commitments of $860,125 in Palm Beach County, $554,000 in Little Rock, and $1,046,500 in New Orleans—a strong indication of continued significant private sector financial support for local LISC programs in spite of highlighted difficulties in all three sites.

Lessons

There are many community development activities in which to engage the private sector. Program experience indicates that the private sector does have a stake in community development and will act on that interest. Private sector interest can be furthered, as well as constructively used, if (1) leading private sector organizations and individuals are made more fully aware of their particular stake in community development and (2) they are purposefully guided so their actions result in private and social benefits. Program efforts can benefit if guidance comes from credible organizations with a track record, for example, LISC.

Using a tangible and credible activity such as real estate development (or other similar activity) as a "deliverable" to organize private sector support can be beneficial. However, to sustain private sector support there is a need to set realistic expectations, manage those expectations over time, and deliver something before too long.

The problems with private sector relationship building appeared to be at least in part caused by unrealistic expectations. This is particularly true with regard to real estate deliverables. Eichler and LISC sold the private sector on concrete deliverables and the use of market incentives to drive the development process. However, as a highly visible measure of performance, when the CDCs failed to deliver on real estate production, that failure was obvious. Furthermore, as a means, real estate production was necessary for further progress, and when it had not been achieved at the critical transition point, it highlighted a significant program failure. Sustained private sector commitment to leadership and organizational development objectives required accomplishment on real estate objectives, but this happened only in Little Rock.

When dealing with a relatively inexperienced and uninformed private sector support community, perhaps emphasis should be placed on enhancing understanding of the difficulties of community development instead of trying to simplify it. More specifically, when gaining private sector support with promises of real estate development, it might be beneficial to "put on the table" that progress will be difficult—that is, it is widely known that CDC real estate efforts take a long time and sometimes do not work out. This is especially true if—as was the case with the demonstration program—real estate development is used as a means as well as an ends and for start-up community development sites. The education itself (and the discomfort associated with being exposed to the realities of the community development process) might make it more difficult up-front to secure private sector commitment, but it might make for more enduring commitment. Greater engagement in regular development activities (e.g., meeting with CDC members in neighborhoods and at project sites) could help private sector organizations and individuals learn about the complexities and difficulties of community development and also help them form more realistic expectations. With a demonstration effort involving much new ground and uncertainty, learning must be built into the program design—private sector supporters must learn (and be prepared to learn) with program evolution.

Lead sponsorship and corporate support of an ambitious community development effort, such as in the three demonstration sites, involves a significant time, financial, and reputational commitment. For private sector commitment to be most effective, it will have to involve more than just money—staff's and senior management's time and energy will have to be committed, as well as planning, technical, and political resources.

The actions of corporate sponsors and supporters will influence outcomes. In community development start-up sites, ongoing engagement of the lead sponsors and key private sector supporters appears to be a critical component of program success. To be effectively engaged will require keeping corporate sponsors informed of progress. To retain their influence, lead sponsors and other corporate supporters should not be reticent to criticize program effort—either of the paid staff or of volunteers.

The actions of all corporate supporters should be congruent with program design and objectives. For example, if one of the program objectives is to engage the private sector in community development (e.g., have private sector actors work with volunteers on real estate), private sector supporters will be obligated to be active. Also, as was the case in the demonstration, if one of the objectives is to put in place accountability, then supporters have to hold volunteers, CDCs, development team staff, and themselves accountable.

In the LISC demonstration, it was initially mainly the responsibility of the local coordinator to actively engage sponsors and other supporters in program-matic effort. This was accomplished with varying degrees of success. There appears to be a benefit from having the engagement of supporters be structured (i.e., well thought out and presented as such in advance). This could include "staging" opportunities for supporters to have positive impacts on the program and to have planned formal and informal contact with volunteers in their neighborhoods. The potential for local coordinators to engage private sector supporters seemed to recede with the hiring of local LISC program directors, a dynamic that was not anticipated in the program design but that, in retrospect, merited attention.

There was also a problem in lack of a consistent message from LISC corporate officers in New York on their priorities and purposes with the demon-stration (this will be highlighted in Chapter 7). For example, some of the perceived changes in priorities of LISC (especially after Michael Eichler left LISC) were misinterpreted in New Orleans as disrespectful and as representing a taking away of local control. In sites such as New Orleans, with a lack of community development knowledge and experience, national LISC needed to send a more consistent and strong message about objectives, including those pertaining to leadership development, organization capacity building, and local control.

In conclusion, program experience suggests that it would be advantageous for community development program staff persons to act under the assumption that private sector supporters of community development (1) can and should be influenced; (2) need and want to be solicited; (3) are often unsure of when and how to intervene, especially if they are relatively inexperienced; and (4) need to be sent consistent and strong messages about program priorities and the expectations for their (own) commitment and contributions.

Notes

1. The program also tried to link the newly formed CDCs with professionals active in the real estate business—developers, contractors, planners, architects, and attorneys. Recruited by the local coordinators and referred to collectively as the technical team, their role was to provide volunteers with technical assistance (at the program's expense) up until the time the CDCs received predevelopment funding from LISC. Findings regarding the technical teams were reported in Chapter 5; this chapter focuses on local private sector corporate executives, most of whom represented organizations that contributed to LISC.

2. The success of this process hinges heavily on making sure that early contacts are positive and that subsequent contacts reinforce positive first impressions. As noted in Chapter 4, the development teams helped CDC volunteers prepare thoroughly before their early meetings with bankers and other development professionals—just as program staff worked with members of the Local Advisory Committee to deepen their understanding of community development and to cultivate reasonable expectations about what the volunteers could accomplish.

3. Bonnie Weaver was the Director of Florida Philanthropy for the MacArthur Foundation. Rebecca Riley, based in the Foundation's Chicago headquarters, was the director of the Community Initiatives Program, of which the Florida program is a part.

4. The John D. and Catherine T. MacArthur Foundation is based in Chicago and concentrates its grant making there. It also has a Palm Beach County office, which operates the Florida Philanthropy Program.

5. Like MacArthur, Worthen and First Commercial made an up-front commitment to make up any difference between what was committed by others and the target set by the national LISC. Curt Bradbury left Worthen Bank when it was bought out by Boatmans. He moved on to an executive position with First Stephens Investment Company in Little Rock and continued to support LISC. Stephens, subsequent to Bradbury's joining the company, became a LISC contributor.

6. The city also had a strong interest in the LISC program for the same reasons. In fact, the idea of inviting LISC to visit, and the invitation itself, originated with Little Rock's highly respected city manager, Tom Dalton. Voters had just defeated a local referendum to use a supplement to the sales tax to fund a new arena in the downtown area, and the city clearly needed a new approach to downtown revitalization. The banks responded immediately, and assumed the lead role to ensure essential private sector support.

7. The Greater New Orleans Foundation did not have the same direct contacts to private sector organizations as the Economic Council, whereas the Little Rock Chamber of Commerce did not have a staff person with the same capability as Dale Smith to assist the local coordinator with strategy.

8. The first chair was succeeded by the mayor of Little Rock in 1995.

9. After two years, members of the strategy committee in Palm Beach County knew many of the neighborhood volunteers they visited by name.

10. The host in Little Rock (the Greater Little Rock Chamber of Commerce) did not seem to play much of a role beyond providing office space. However, it did give the program and development team staff a credible home in the business community.

CHAPTER 7

Transition and Its Consequences

Objectives and Strategy

The demonstration program was designed to be of short duration. LISC raised enough funding at the outset to support the development team's staff and activities for a specific time period (two years in Palm Beach County and Little Rock, three in New Orleans). The staff's mandate was to manage the process of organizing and strengthening the CDCs and the support community and to foster positive relations (bridges) between the two so that community development efforts would function effectively without the development team—but with the core LISC program administered by an on-site program officer or program director—when the program concluded. Simply put, the development team was supposed to work itself out of a job and transfer many of its responsibilities to volunteers, the CDCs, LISC local program directors, and leaders in the local area support community.

The finite character of the organizing program had appeal to all the major players at program start-up. Private sector supporters, particularly those unfamiliar with community development, liked the idea of a program that delivered a product on a known and relatively short timetable. They could then look at the results and decide whether what had been accomplished merited their continued support.

However, the compelling rationale for a finite intervention with a known deadline came from Michael Eichler. He argued that a clear deadline provided

125

volunteers were to be prepared for this new responsibility. Equally important was the fact that once the CDCs assumed responsibility for the coalition, the development team no longer had a role; the volunteers were in control, and external agents were not to interfere unless asked for help or advice.[3]

Local coordinators had primary responsibility for managing the transition. This included ensuring that the organizers prepared the volunteers on each CDC board to participate in the process of deciding whether to form a coalition, coordinating program activities with national and local LISC staff, and pacing program progress to ensure that local community development activities did not lose momentum when the development team departed and another arrangement took its place. The key tension in managing the program's pace was to avoid having the development team leave either too early (before sufficient local leadership and capacity was established) or too late (after dependency on outsiders, i.e., the development team, was established).

Positioning the volunteers to assume responsibility for forming (or not forming) a coalition required bringing volunteers from different CDCs together, both informally and formally, helping them structure a decision-making process (both collectively and in their respective CDCs), providing them with information, and coaching them through the decision-making process. During most of the program, contacts among volunteers from different CDCs were largely informal, for example, at social events or after encouragement from the organizers to get to know volunteers in other CDCs who were dealing with similar issues.[4] About six months prior to the planned transition date, the coordinator (working through the organizers) encouraged the CDCs to set up a series of structured meetings to discuss the coalition issue; this included exposing volunteers to the benefits of organizing a coalition (e.g., having volunteers in Palm Beach County and Little Rock meet with visiting delegates from MVI), and, to varying degrees in the three sites, encouraging the volunteers to organize coalition organizations in much the same way they encouraged them to organize CDCs.

At the level of the support community, transition-related activities were generally less structured. The LISC Local Advisory Committee received quarterly reports on the CDCs' progress on their development projects and was informed when the CDCs began formally to consider the coalition decision. Funders also had an opportunity to meet with delegates from the MVI when they visited the sites. However, informal (but regular) conversations with members of the support community, especially key funders, were the principal way program staff sought to educate them about the benefits of a coalition and build their support for the idea. These conversations took place mainly as the planned transition date neared; their timing and character were intended to ensure that funders were prepared to embrace and support the decision the coordinators anticipated the volunteers would make.

Findings From the
Implementation Experience

Efforts to leave in place a coalition of CDCs were ultimately unsuccessful in all three sites. In both Palm Beach County and Little Rock, coalitions were established and then experienced difficulties. In Palm Beach, the process of choosing an executive director was difficult and divisive. In both places, the executive directors hired were not right for the position and problems ensued. These problems ultimately led to diminished support for the coalition by both funders and volunteers, disagreement between LISC and the Consensus Organizing Institute (COI),[5] and finally, the dismantling of the coalitions. This outcome was the product of poor program design regarding the management of the transition, a series of implementation shortcomings, and disagreement about priorities between LISC and COI. In New Orleans, a coalition was never formally organized.

The Deadline for Transition

The original plans were to terminate the development team's work after two years in Palm Beach County and Little Rock, and after three years in New Orleans. Only in Little Rock was the target date met; in the other two sites the length of the demonstration program was extended.

In Palm Beach County, it was clear months before the planned conclusion of the demonstration program that the CDCs would not have made sufficient progress on their real estate projects by the end of the two-year period. Program staff and funders agreed that the life of the development team should be extended so that the CDCs would have a better opportunity to demonstrate their competency before funders were asked to make commitments about providing long-term support. The Local Advisory Committee approved an 18-month program extension, and LISC raised enough money to support the development team for that period.

The need for an extension in Palm Beach County was not a major issue for program supporters. They understood that this was the demonstration's first site, and that the initial timetable had been very ambitious. They were disappointed, of course, especially those whose interest in the program was motivated mainly by the promise of real estate projects. Nevertheless, key funders remained strongly committed to the program, wanted it to succeed, and believed that the chances were still good that it could work. All agreed, however, that if, for whatever reason, a second extension were to be necessary, the program would be widely perceived as being in serious trouble, and that raising funds to support a second extension would be difficult.[6] They therefore approved a relatively long

(2) individual CDCs that would each have their own staff and would work independently. The former option appeared always to be presented as the better choice by Eichler and development team staff.[13]

The benefits of sharing staff, sharing information and knowledge, and joining together to influence city and county community development policy were the main "selling points" when the coalition option was discussed. However, in both Palm Beach County and Little Rock, volunteers were also strongly influenced by their interaction with the executive director of MVI and a group of MVI delegates, who visited the sites while the CDCs were considering their options. In contrast, the volunteers were told about the experience in Houston (LISC's initial test of Eichler's consensus organizing approach, after the Mon Valley pilot program), where the CDCs decided not to form a coalition and some of them did not survive.

Nevertheless, CDC board members brought their own issues and perspectives to the discussion and the decision. In supporting the formation of a coalition, many board members valued the notion that the CDCs would build bridges and bond. For African American volunteers, in particular, this sometimes seemed also to have the connotation of the importance of the black community "sticking together."

The strongest reservations about the coalition approach were voiced by one or two CDCs in each site that had a strong inclination to remain on their own. Their entrepreneurial leaders, particularly those who had some prior experience working on community issues, felt they could do better without a coalition—for example, raise funds to support their organizations—and wanted direct access to local funders. In Palm Beach County, these leaders were individuals who were accustomed to positions of authority and were confident of their ability to lead "their" organizations well. Ultimately, however, they were persuaded to give the coalition a try—in part because they were sensitive to the issue that if the CDCs tried to compete for funds some might fail, and in part because they understood the coalition's potential to gain greater political influence for the county's communities of color.[14] In Little Rock, one CDC made an effort to obtain a grant from the Levi Strauss Foundation (a LISC contributor) and was told that the foundation was interested only in supporting a coalition. This undercut the possibility of remaining independent and left some of the CDC's volunteers feeling that the coalition was a "done deal" because the funders had already made up their minds.

In all three sites, several important issues received little or no explicit consideration. The most critical of these was that the CDC volunteers were not adequately prepared (on a number of important dimensions) to take on the new responsibilities associated with governance of a coalition. The local coordinators—who were, after all, doing this for the first time—were operating with little guidance from either the program model or Eichler; they were in the very challenging position of trying to work themselves out of a job (and, in two cases, to leave town) on a tight timetable, and they were getting little help from either core LISC program staff or local program supporters.

The Coalition Experience

The demonstration program assumed that volunteers' experience in working in their CDCs, combined with the process of deciding to form a coalition and getting it initially in place, would prepare them to oversee and guide the organization of the new entity.[15] It did not, however, primarily because the coalitions were much more difficult organizations to govern and manage than the CDCs. This greater difficulty had at least four dimensions.

First, CDC-related work was neighborhood oriented; its purpose was clear to board members, and it was concerned with issues with which they were familiar. The coalition's mission was both more sophisticated and less clear to the volunteers. Operating the coalition required volunteers to learn and think more about each other's organizations and activities, about decision making in a more diverse and politically complicated group, and about "bigger picture" community development issues (e.g., shaping city and county community development policies) and political relationships. Organizing an effective coalition thus required a more sophisticated understanding of the community development process and of local politics than many delegates had, including an appreciation of the complementary nature of coalition and CDC efforts.

Second, CDC-related work was more structured. For most CDCs, that work had centered on real estate development—a process with a clear structure that had been laid out for board members in a clearly organized step-by-step fashion. In contrast, the exact purpose of a coalition (and, hence, the content of its work) was less clear for many, perhaps most, volunteers. Beyond the need to hire an executive director, there was no obvious task. This was to a large extent unavoidable because setting the coalition's agenda was, by intention, the task of the CDCs' delegates to the coalition board—that is the meaning of volunteer control—but they needed help working through this difficult process as a newly organized group.

Third, CDC board members had little experience hiring and supervising staff. Some coalition delegates had jobs that gave them experience hiring personnel, and the local coordinators did try to explain to delegates what their task entailed, but this was weak grounding for the critically important task of selecting and overseeing an executive director of a coalition organization.

Finally, as described in Chapter 5, the development teams made cultivating trusting and supportive relationships with the CDC volunteers a priority. Board members working for their CDCs had the benefit of the assistance and support of their community organizer, who could in turn provide them with access to other helpful people. At transition, however, the board members' responsibilities increased substantially (for the reasons discussed previously), at exactly the same time that their relationship with development team staff was ending.

The new coalitions in Little Rock and Palm Beach County, just like the newly organized CDCs that preceded them, would have benefited greatly from more extensive organizational development and technical assistance support. These types of assistance are commonly provided in a variety of formats, and it is not clear from this research exactly what venue would have been most desirable.[16] What is clear is that formal and ongoing assistance would have been helpful, that transition assistance should have been initiated prior to the hiring of an executive director, and that, ideally, initial assistance should have been focused on helping delegates hire a competent and committed executive director and structure an initial stream of work (e.g., lay out tasks and products to be completed on an agreed-on timetable). In general, the approach that would seem to be most consistent with the overall program strategy would be one in which delegates would be given the support they needed to "get up to speed" and would then engage regularly with members of the support community in a way that would strengthen the understanding and shared commitment of both.

Program experience illustrates yet again that capable leadership is critical to the success of fledgling organizations. In this instance, hiring talented executive directors was key to the ultimate success of the coalitions. This was difficult in Palm Beach County and Little Rock. The job is challenging and requires a combination of substantive knowledge (about community development and management), interpersonal skills, and political sophistication. In addition, it would be helpful for the individual hired to have some specific ties to the area, CDCs, and volunteers. In areas with a paucity of prior community development activity, identifying and recruiting someone with all these key attributes is almost certain to be a struggle—and volunteers must be in a position to understand the importance of these traits and assess their strength in prospective candidates.

The local coordinators in Little Rock and Palm Beach County took different approaches to the task of helping volunteers select an executive director. Each proved to have its pitfalls, and each was ultimately affected adversely by local attitudes about race and by the impending deadline.

In Little Rock, Richard Barrera used a process similar to the one Eichler had used to recruit both the local coordinators in the demonstration and the first executive director of MVI. Barrera had developed an extensive network of contacts over the course of the demonstration program. His experience with them gave him a good picture of the local talent pool, and he used his observations and conversations with them to assess their suitability for the executive director position. He spoke with volunteers and members of the support community to determine how they felt about individuals who might become candidates for the position, trying to educate them in the process about the qualities the executive director would need. When he felt he had a good idea of who the best candidate would be, he also discussed the matter with Eichler.

Barrera's support for Willie Jones (the organizer who was eventually hired as the executive director) was based on a number of factors including Jones's understanding and commitment to the consensus organizing approach; his potential for further growth (although he had limited technical development skills, he had grown a lot as an organizer, and Barrera believed he had the potential for further growth); his positive relationship with volunteers (he had gained their trust during his work as an organizer); and his being an African American—a fact that both Barrera and many volunteers thought was appropriate given the primarily African American composition of the CDC boards and the civil rights history and climate of Little Rock.[17]

Unfortunately, with the abrupt departure of Richard Barrera, Jones was left on his own to "grow into the job," and he didn't. He did not see it as his job to reach out to the support community. They, in turn, did not see it as their responsibility to give him unsolicited advice or initiate discussions of strategy on program activities; they waited for him to come to them—as others, including Barrera, did—and he lost their support (as discussed in Chapter 6). A large share of the funding for the coalition in Little Rock was provided by two private foundations—the Levi Strauss and Winthrop Rockefeller Foundations; LISC did not provide any operating funding for the coalition. The two foundations did not provide any special mentoring support to the executive director or the coalition board, as their expectation was that others—for example, LISC staff, and members of LISC Local Advisory Committee and COI (including Richard Barrera, who had joined COI soon after leaving Little Rock)—would do so.

The CDCs and their delegates to the coalition never really internalized the idea that the coalition was supposed to work for them and that their responsibility as delegates was to assure that it did so. When, over time, it didn't do so, their commitment to the coalition and their support for Jones waned. When the funders, with the concurrence of the local LISC program officer, declined to give the coalition further support, it had no real constituency among the CDCs.

Mary Ohmer in Palm Beach County began with a process much like Barrera's, but the virtual collapse of the Palm Beach County technical team meant that the pool of seasoned professionals who had experience with the program was much smaller, and it included no suitable candidates. Likewise, no one on the development team was right for the job. She fell back on a more traditional, formal search process in which she essentially acted as staff for the hiring committee formed by the coalition delegates. The hiring committee in Palm Beach County drafted a job description and placed job advertisements in local papers, in selected major metropolitan newspapers, and in the appropriate publications of relevant professional associations. Ohmer also called contacts elsewhere in the country to see if they could suggest candidates.

The applicant pool in Palm Beach County was narrowed to three individuals, including one African American from outside the area and two local candidates, one African American and one white. The candidates were screened by an interview committee composed of the coalition's hiring committee and the incoming and outgoing chairs of the Local Advisory Committee. Making a choice from the shortlist proved difficult and, in the end, deeply divisive. The Local Advisory Committee members thought the one candidate from out-of-town was not qualified, but that either of the others would be fine. Ohmer agreed with them, although she had a clear first choice. Unfortunately, when asked for their views, the advisory committee members (both white) were uncomfortable speaking against an African American candidate; instead, they said that any of the candidates would be acceptable.[18]

Left without the support of the advisory committee members, Ohmer's efforts to build support for her preferred candidate and to point out the signals that the external candidate would not be a good choice backfired. Many volunteers saw them as a heavy-handed effort to manipulate the outcome while assuring them that the choice was really theirs; in their eyes, the program was not behaving consistently with its articulated principles. After an extended process and several rounds of voting by the full board of delegates, a closely divided vote selected the out-of-town candidate, who accepted the job.

Two immediate factors shaped this outcome; one was race. Early in its deliberations, the committee was close to recommending the only white candidate when one of the African American delegates voiced the view that the executive director "had to be black." White delegates were indignant, pointing out that color was never discussed as a job qualification—but once race had been made an explicit issue, many African American delegates found it progressively more difficult to support a white candidate, and the third candidate (an African American) could not sustain adequate support. At the same time, some white delegates were reluctant to support the white candidate they favored because they feared the impact on the organization of a final selection polarized along racial lines.

The second factor was time. Everyone knew that the development team's deadline was fast approaching. Ohmer took on the "committee staff" role in part because it would make the process move more quickly than following the program's normal pattern of having the volunteers do as much of the work as possible. Unfortunately, the fact that she became the funnel for information about the candidates (e.g., doing all the reference checks) contributed to the delegate's perceptions that she was trying to manipulate them. If the context had allowed more flexibility, interviewing additional candidates or taking steps to expand the pool would have been an attractive option, but the impression that there wasn't time precluded this. And the fact that the delegates had not had an opportunity to develop a successful track record of working together (as the volunteers originally had when they formed their CDCs) meant they had no reservoir of shared experience and goodwill to draw on when the issues on the table became both difficult and racially charged.

The new executive director's conduct on the job exacerbated rather than ameliorated the problems at the site, including the divisions among the delegates, the weakened CDC boards, and the absence of real estate production by four of the six CDCs. The Local Advisory Committee finally signaled a vote of no confidence to the volunteers; the executive director departed, and after a lengthy delay, a new local LISC program officer began the task of stabilizing the county as a more typical LISC area of concentration, at which LISC provides CDCs with modest operating funds to support staff.

One additional factor tended to weaken volunteer support for the coalition idea over time. Most volunteers felt greater comfort with and commitment to their communities and their CDCs than to a larger coalition effort. For example, in all three sites—but most notably in Little Rock—CDC board members wanted their organization to have a presence in their neighborhood as a visible sign that the CDC was working for the community with an office residents could call or visit. Volunteer loyalty to the CDCs also affected the choice of coalition delegates in a way that had negative implications for the coalition in Little Rock and discussions on organizing a coalition in New Orleans. In many cases, either the best volunteers choose not to be delegates or the CDCs elected to send as delegates board members whose primary energies they felt the board could afford to lose.

The Role of the National Partners: LISC and COI

During the course of the demonstration program, Michael Eichler left LISC to form a new nonprofit organization: the Consensus Organizing Institute (COI). Eichler had aspired to establish his own entity for some time, but the timing of the decision by the two parties influenced, and was influenced by, the demonstration program—particularly the lack of progress on real estate production by most of the CDCs in Palm Beach County and New Orleans and disagreement over the causes and appropriate next steps. When Eichler left LISC and was joined by local coordinators at COI, COI assumed responsibility for overseeing the development teams under a short-term management contract with LISC.

In the original program design, the goals and priorities of LISC and COI (then actually Eichler and the development teams, officially part of LISC) were shared; their skills and emphases differed, but in ways that were viewed as mutually beneficial. LISC's experience showed that meeting national quality standards for affordable housing production would help CDCs develop organizational capacity, establish credibility, and thereby gain access to additional resources and skills. Eichler's experience demonstrated that volunteer engagement, commitment, and control ensured that resources would be used to pursue the community's agenda and meaningful voice on community development issues. The premise of the program was that these two perspectives were complementary and would strengthen one another.

As the program unfolded and the development team process appeared not to be leading to the expected level of real estate production in two of the sites, their distinct emphases became more apparent.[19] If the program could not have the best of both worlds, each partner gave primacy to the perspective that had brought them to the table (and on which each felt that their reputations depended). For LISC staff and many LISC Local Advisory Committee members, the main objective was for the CDCs to operate in a "business-like" manner and to effectively produce housing. For Eichler, the development team staffs under his management, and later for COI, the priority was to develop neighborhood leadership and have them put in place a coalition organization to promote local volunteer control of community development. The tension between the two priorities was not an issue in Little Rock, both because housing production went well and because the Little Rock Local Advisory Committee was more concerned than its counterparts in the other sites with neighborhood leadership development and capacity building.

Divergent priorities, each grounded in both values and experience, made good communication and concerted action difficult. When the primary national program partners' communication abated, the difficult management of tensions between resident control and production levels in the sites received less attention from national staff than it required. This put greater pressure on local staff: local coordinators working their way out of a job and LISC national and local staff who had problems of their own (in Palm Beach County and New Orleans because production was behind schedule, and in Little Rock because there were two large National Equity Fund projects in progress, as well as activities with all the CDCs).

The experience in New Orleans highlights the problems of not having efforts between national community development partners (in this case, LISC and COI) and among program participants well coordinated, especially at critical junctures when the allocation of roles and responsibilities is in transition. Local participants in New Orleans were uncertain as to who was in charge—was it the departing local coordinator, Michael Eichler, the local LISC program director or national LISC staff, the chair of the LISC Local Advisory Committee, volunteers? This was detrimental not only to a smooth transition, but also to volunteer control and credibility, because real estate production—already well behind schedule—suffered when volunteers were uncertain who to call for assistance (the local coordinator or local LISC program director?). It also contributed to an overall loss of program credibility and a decline of confidence among major funders.

To many of the local program participants, there was also the problem in New Orleans of national LISC senior staff failing to provide consistent and constructive input. Program participants in New Orleans felt disconnected to LISC national staff. As a result, periodic suggestions from national LISC were interpreted by some local actors as disrespectful to local culture and control and generated some defensive, instead of constructive, reactions.

In Palm Beach County, differences of opinion about who was responsible for what (differences initially shared by the funders) meant unfortunate delays in addressing the situation "on the ground" that was deteriorating quickly. The result was wounds that will take some time to heal.

This raises directly the issue of accountability; to whom were COI and LISC ultimately accountable? Across the sites there were lessons about the importance of making outside agents more accountable to local parties, especially funders on the LISC Local Advisory Committee and the target population. As accountability was questioned, there was a decline in support-community commitment to local programs when they perceived less influence and fewer results from their participation. This perception among funders was strongest in New Orleans but also relevant in Palm Beach. In New Orleans, over time local funders became increasingly concerned that the program was out of their control and not accountable to local actors (e.g., "we put up $1 million locally and LISC in New York hires and fires the local program director—that's not right"). This affected program credibility and the confidence that funders had in LISC locally.

Program experience suggests that greater accountability could have been achieved with more frequent and substantive communication throughout the program. Better communication could have included regular status reports on where things were with regard to achieving the necessary conditions for effective transition—including progress on CDC organizational development and activities other than real estate production, and consultations (i.e., coordinators seeking advice from members of the support community and using those conversations to help shape the support community's expectations).

The experience of the local coordinator in Little Rock illuminates this point. Although he did not organize a formal strategy committee, he effectively engaged in ongoing conversations with major funders and members of the support community. These conversations intensified at and helped transition. The experience in Little Rock suggests that more frequent and fluid communication can provide a vehicle for outside agents (e.g., coordinators and LISC staff) to, over time, transfer knowledge and ultimately responsibility to local agents. In Little Rock, when the local coordinator left, the coalition's executive director failed to keep communication channels active and this contributed to the coalition's shortcomings.

Program experience also highlights the difficulty of initiating efforts in areas such as the demonstration sites. The program consistently underestimated the degree of difficulty and the time necessary to realize benefits. Volunteers at transition were not ready in any of the sites to assume control to the degree envisioned, especially in a coalition context, and the supportive bridges and infrastructure they needed were not in place. The experience in the Mon Valley with MVI suggests that hiring an exceptional executive director can help overcome this shortcoming (R. Gittell 1992); however, that proved to be difficult in Palm Beach County and Little Rock.

In conclusion, the objectives behind the strategy to exit with a coalition organization in place were neither unimportant nor unworthy; however, the transition strategy itself was neither well conceived nor effectively implemented.

Lessons

In the context of an intervention strategy explicitly designed to be of short duration, transition is a critical issue. The transition period is the most visible aspect of most interventions, both for national and local observers, and it is a natural point for observers (e.g., the authors) and participants to assess what has been accomplished. The transition strategy therefore needs to be as carefully thought out, articulated, and sensitive to the concerns of the program's varied constituencies as the start-up of the initial program.

In the context of a program as complex as this one (and it was much less complex than many community development initiatives started in recent years, as highlighted in Chapter 3), developing such a strategy entails thinking of transition as a *process*. Program transition may appear to be an *event* (e.g., "now the CDCs have a coalition and are in charge"), but the program design and strategy must reflect the sustained effort that lies hidden behind an effective "event." A solid program transition requires laying the ground work, learning and practicing the skills, teamwork, good delivery, careful timing, and follow-through.

The character of follow-through needed will depend on—among many other things—the demands that the post-transition program places on participants. In this instance, those demands increased. Program experience indicates strongly that under such circumstances, support for participants needs to ratchet up accordingly. The fact that participants have made the grade during one phase of the work does not ensure that they are positioned to accomplish substantially more difficult tasks successfully. However, the types of assistance provided—and, importantly, how assistance is offered and provided—will likely have to change to reflect participants' increased capabilities and position. If this cannot be accomplished (or, indeed, even if it is accomplished), the wisdom of having a back-up plan (the programmatic equivalent of an insurance policy) is again apparent.

Program experience at transition highlights the difficulty and delicacy of balancing outside direction with local control. In each site, the development team staff, toward the end of their tenure, challenged both local volunteers and private sector supporters to assume greater control of, and responsibility for, local community development. At the same time, they sometimes inadvertently sent contradictory signals to both—to volunteers who were just starting to recognize and act on their new power and to Local Advisory Committee members who, on the whole, would have preferred to continue to have LISC staff (meaning both the LISC local program director and development team staff) responsible for making the program run smoothly. A shared, well-informed understanding of what balance is desired (between local and outside responsibility) would appear to be a key to making any balance work.

The tension between *process* and *product,* discussed at several points, confirms that greater attention is needed to the issue of how to define, measure, and report on community development program performance. The logical steps of the real estate development process provided more detailed indicators of "hard" progress than is often available, and Local Advisory Committee members heard reports on them. But the development team also had hard indicators of progress other than real estate (indicators they used to gauge both organizer performance and how well the program was doing) that could have helped the support community develop a more sophisticated understanding of community development's requirements and opportunities. Regular and close engagement by program supporters with volunteers and program staff (e.g., visiting with volunteers at development sites in neighborhoods) could have enabled supporters to understand and consider broader measures of program performance, including progress on such intangibles as enhancing volunteer knowledge of community development, increasing volunteer commitment to their neighborhoods and CDCs, and neighborhood leadership development. These types of intangibles are integral to any complex community development intervention, but cultivating an appreciation of them will, in most settings, require an investment. How great an investment any program and its supporters will find it worthwhile to make will be related to how they think about the distribution of responsibility, authority, and accountability.

Notes

1. See Chapter 1 for a more detailed description of the Mon Valley Initiative.

2. Good progress on real estate projects by all (or most) of the participating CDCs also entered into the dynamics of making a decision about forming a coalition, because this was a clear way to position the CDCs to enter into a collaborative relationship as equals.

3. This aspect of the program derives from a central tenet of community organizing that a legitimate community organizing activity requires a *mandate* from the community. Eichler believed

strongly in this principle, which was also central to the neighborhood selection process: Residents had to *want* the program for the neighborhood to be chosen.

4. One notable exception to this pattern in Palm Beach County is that CDC presidents worked jointly to press for changes in county funding policies.

5. Eichler left LISC in 1994 and organized COI. COI then assumed management responsibility for development teams under contract with LISC.

6. Indeed, the program's skeptics, including county civil servants responsible for housing and community development, took the lack of housing production and the need for any extension as evidence that the program was ill conceived and in trouble.

7. Recall from Chapter 5 that by this point in the program, the weakness of the technical support team had taken its toll on project progress, diverted the organizer's time and attention to the real estate projects so that maintaining the organizational strength of the CDCs got short shrift, and left development team staff overextended.

8. Eichler and Ohmer had ample notice that Tarr would have to be replaced and did a serious, although informal, search (working through their extended networks, as they had done at program start-up) while Tarr was still with the team.

9. Eichler did not think that the organizer, Willie Jones, was the right person for the job and tried to persuade Barrera to find and support another candidate. Barrera did consider others, but ultimately supported Jones. Eichler did not interfere—an example of his management style of giving staff latitude to make what he might consider mistakes and then learn from them.

10. Delegates did begin to meet in New Orleans, however, and had elected officers and begun the work of making decisions about articles of incorporation.

11. As noted previously, during most of the demonstration program, volunteers focused their energies on the tasks and issues facing their own CDCs and neighborhoods. Contacts among volunteers from different CDCs had been largely informal. For example, the coordinator in Palm Beach County encouraged the volunteers to have a program-wide Christmas party the first year the program was in operation. However, it was not unusual for volunteers from different CDCs to know one another (or at least know of one another) through their participation in other activities.

12. The local coordinators in Palm Beach County and New Orleans, Mary Ohmer and Reggie Harley, thought from the outset that a coalition was the preferred alternative. The Little Rock coordinator, Richard Barrera, was initially unsure whether that was the best option, but as the program progressed he became convinced that it was.

13. The fact that only two alternatives were put on the table reflects the experience of the key people collaborating to design and deliver the program: Michael Eichler and Richard Manson. Each had personal experience with both of these program models. LISC had worked with both models but much more frequently with individually staffed CDCs and was thus more comfortable with this option. No one involved in the program, including funders and volunteers, seriously raised the issue of developing a third alternative until the funders in New Orleans perceived that the program was in trouble and were uncomfortable with the options offered.

14. The volunteers in Palm Beach County had a better understanding of this issue than their counterparts in the other sites because of the demonstration program's difficulties in dealing with two key county officials prior to the formation of the coalition. All the CDCs needed some type of subsidy dollars to make the housing they sought to develop affordable to local residents; for four of the six CDCs, the county was the key source of public subsidies. However, county policies governing the allocation of federal CDBG and HOME funds made the CDCs' proposed projects ineligible. Collective action by the six CDC presidents played a critical part in getting the county regulations changed and hence in the CDCs' access to important resources. This gave CDC board members a clear demonstration of the value of working together on issues of common concern. As a result, the more politically savvy volunteers, particularly the presidents, brought to the coalition decision a sharper appreciation of the larger political value a coalition might provide.

15. The process of putting the coalition in place included not only the decision-making process, but also agreeing on a name, mission statement and structure, electing officers, and hiring an executive director. Agreeing on a structure was straightforward because both sites had MVI's articles of incorporation and by-laws and used them as models.

16. For example, formal training might have included evening meetings or all-day retreats that mixed presentations with workshop sessions. Ongoing support could have taken such forms as linking delegates to mentors experienced in community development and organizational management, or adding "seasoned" members from the Local Advisory Committee (or other knowledgeable supporters) to the board as nonvoting members or advisers.

17. The second strongest contender for the job was a white woman who had been a member of the technical team and was the attorney for one of the CDCs. She, too, was committed and well liked (although known well by fewer volunteers), and she had a much stronger educational and technical background.

18. See Chapter 6 for a fuller discussion of the program's difficulties in cultivating among private sector supporters a real understanding of the implications for their own behavior of allowing the volunteers to assume control. Supporters commonly found it difficult to come to the notion that volunteer control was not inconsistent with a "business-like" airing of divergent points of view, particularly across lines of color and class.

19. Recall from Chapter 5 the tensions between process and product with which the CDCs grappled in all three sites—a microcosm and, in a sense, a foreshadowing of the tensions between the partners.

Lessons

Building Social Capital

The LISC demonstration program was unusual not only because it sought to test the broad applicability of a new approach to community organizing, but also because it attempted to use community organizing as a strategy for promoting community development. This novelty, combined with the clarity of the program design and the richness of the program's experience, also makes it a good vehicle for learning about community development interventions.

This chapter presents lessons in two parts. The first part focuses specifically on the LISC demonstration program, distilling the key program attributes that shaped its performance in pursuing its central objectives and identifying the lessons it has to teach. The second part uses the analysis of the program's design and implementation to reposition the program in its context, both programmatically and theoretically, and to reflect more broadly on what its experience tells us about those contexts.

Key Aspects of Successful
Community Development Interventions:
Lessons From the LISC Demonstration Program

The main objectives of the LISC demonstration program were progress on commitment, capacity, and control (what we refer to as the big "C's") among neighborhood volunteers, members of the support community, and staff.[1] Adequate progress toward these interim outcomes by the time the demonstration concluded constitutes grounds for optimism that the program created the basis for sustainable community development in the sites.

The research has identified 11 key aspects (the little "c's") of programmatic effort that strongly affected outcomes (i.e., performance on the big C's) and sustainability (see Figure 8.1). These are the factors that consistently had an influence on progress toward the three major program objectives. Six of these are intermediate outcomes (as defined in Chapter 2): *comprehension* of community development; *credibility* of the program and its participants; *confidence; competence; comfort* with the goals and objectives of the program and working with other participants; and constructive *critiques* of program progress. In general, these were absent at the outset of the program because the demonstration sites, by design, lacked previous community development experience.

The other five key factors that shaped program outcomes are aspects of program implementation: *communication* among participants; *consistency* of purpose; *congruence* of program efforts; *counterbalancing* inherent program tensions; and adjusting implementation to reflect local *context*. The three sites had different experiences with regard to these components, and as suggested by the activity analysis and summary discussion here, this seems to explain much of the variation in outcomes across the sites.

As the following discussion indicates, commitment was most affected by confidence, comprehension, credibility, and congruence; capacity was most affected by comprehension, competence, critiques, communication, and consistency; and control was most affected by confidence, comfort, and comprehension. All of these (big C's and little c's) were affected by how well program staff and key sponsors counterbalanced the tensions inherent in any externally driven community development effort and adapted implementation to suit the local context.

The next two sections of the chapter explore, in turn, each of the key dimensions of the program that influenced its performance. For each, we begin with a brief definition, then cite illustrative examples from the discussion of program experience in Chapters 4 through 7, and conclude with lessons drawn from our analysis of that experience. We describe the intermediate outcomes first, then program implementation issues. We undertake this discussion with

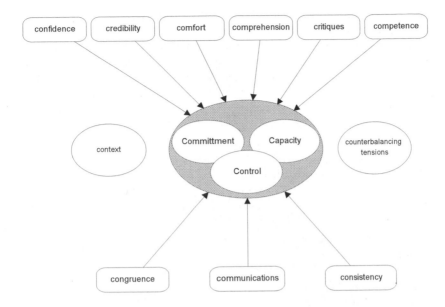

Figure 8.1. Key Elements of Sustainable Community Development:The Big and Little "Cs"

the caveat that these factors are highly interrelated and thus difficult to separate analytically.

Key Intermediate Demonstration Program Outcomes

The six intermediate outcomes were both program goals in their own right and stepping-stones to a sustainable local community development system and the benefits it can generate—that is, each is both a means and an end. The program sought to establish each of them, and if community development has really taken root in these localities, each will grow stronger over time as the system matures.

■ Comprehension

Comprehension refers to *system knowledge,*[2] that is, how well program participants understand the nature of community development processes and the effect program strategies and tactics are expected to foster in them. It also includes how well participants understand their own interests, roles, and responsibilities and those of other community development participants. Acquiring this deep understanding and sense of perspective is central to how program participants develop trust in one another, based on an understanding of their mutual interests and a shared vision of the program and its potential—three of the four dimensions of social capital identified by Keyes et al. (1996). Comprehension has an important effect on both community development commitment and capacity.

Comprehension of community development clearly increased in all three sites. It grew most sharply in the early stages of the demonstration, but transition-related issues also generated significant new learning for many participants. Between those two points, participant understanding grew, but at a slower pace. New Orleans started from the lowest initial level of comprehension and improved, but at the end of the demonstration, comprehension there was still generally below the level achieved in the other two sites.

In all three sites, site assessment and fund-raising activities raised comprehension in the support community, as did early meetings of the LISC Local Advisory Committee. Initial organizing efforts produced similar gains among volunteers in the targeted neighborhoods, as they thought about their goals, about the varied interests within the community, and about choosing an initial project from among those suggested at the town meetings. As comprehension grew, so did commitment to the program.

Once the CDCs were organized and settled into working on their projects, however, momentum on comprehension trailed off as the primary focus shifted to dissemination of specific, generally technical, information related to project development and finance. Both volunteers and funders clearly needed this information to meet their responsibilities for moving the program forward. Unfortunately, as development and related organizational issues required increasing levels of effort, staff attention to expanding comprehension of broader community development issues among supporters generally declined and became rather uneven. As a result (and absent a climate in which collectively "taking stock" from time to time was the norm), the LISC Local Advisory Committees generally did not acquire a good understanding of how to manage the tension between supporting the CDCs (and their coalitions in Little Rock and Palm Beach County) and holding them accountable for performance and quality.

In Little Rock and Palm Beach, activities surrounding the formation of coalitions, including site visits from members of the Mon Valley Initiative, fostered comprehension by focusing the attention of volunteers and funders on long-term, big-picture issues. However, in both places, relatively few volunteers or funders fully appreciated either the potential benefits of the coalition approach or the potential difficulties it might pose.[3]

Inadequate attention to comprehension appears to have had its strongest negative consequence in New Orleans, where program efforts at transition were refocused on real estate development because progress had been minimal. Because the program did not produce the interim outputs funders had been led to expect, it was ill positioned to press to the top of the agenda such big-picture issues as whether to support a CDC coalition in New Orleans, or whether progress had been made on some of the issues that had made the demonstration program attractive in the first place; these included how the city engaged in community development (i.e., through patronage or a merit system) and how the Greater New Orleans Foundation could more effectively work with private corporations in community development. The lack of continued progress on comprehension among supporters appeared also to be debilitating to volunteers who, particularly after significant delays with real estate projects, came to feel that they had limited control over real estate efforts, and thus limited potential for future influence over their community's development.

Lessons. Program experience suggests that there are long-term advantages to paying ongoing attention to increasing general comprehension of community development. In places with a well-developed cadre of CDCs, participants at both the grassroots and support levels reflect on their practice and take stock of how they must adapt to changing circumstances and lessons learned. That is one of the reasons they remain at the cutting edge of the field. Sites where community development is relatively new have even more to gain from setting aside time to process and share what they are learning and to seek out approaches that have worked well elsewhere and might have local applicability. Initiatives that are part of a national network (in this case, LISC's network) have a built-in learning opportunity that should be used to advantage. In the demonstration program, key individuals thought hard about their local program (both what was going well and what needed to be strengthened) and often had very interesting insights, but in none of the sites did the program create opportunities for groups of participants to do this collectively.

Structured opportunities to reflect and learn will yield lessons about program substance—real estate development issues, in this case. However, they are also likely to present chances to broaden the discussion to larger, more difficult issues, such as politics, power, and race. Managing such conversations well is always a challenge, but could yield important payoffs if they clearly have roots in the program and implications for its success. Furthermore, periodically reconnecting with some of the more fundamental, underlying issues and goals of an intervention can serve to motivate and energize those participants—both volunteers and supporters—whose interest in technical issues is limited but who have a broad commitment to community development and want to change communities in fundamental ways. Increased opportunities for communication among various actors could also help to enhance comprehension, particularly related to the interests, incentives, and actions of others.

Active consideration of alternative ways in which comprehension can be fostered is likely to be useful. In the LISC demonstration, the two national program managers and the local coordinators (plus the LISC program director in New Orleans) bore the main responsibility for promoting comprehension, but the organizers were the main point of contact between the program and the volunteers. Most of them had very limited knowledge of broader community development issues at the start of the program, but it was well within their grasp; they learned quite a bit and could have learned more. Further expansion of the organizers' comprehension of community development throughout the program would have been beneficial. Organizers could have then transferred greater knowledge to volunteers. An additional benefit would have been that the organizers' credibility would have been easier to sustain if they had continued to bring important new knowledge to resident volunteers.

■ Credibility

The credibility of the program—its reputation, as well as local staff and sponsor reputations—had a significant effect on outcomes. Its most direct impact was on the commitment of the support community and the volunteers.

Credibility in the Support Community. The demonstration program initially enjoyed very high credibility with members of the support community in all three sites. This "in-the-door" success was a product of LISC's national track record and reputation, the high personal credibility of the two LISC managers responsible for site assessment and overall program management (Eichler and Manson), and the prominence and influence of the local sponsors.

Once the program was under way, the responsibility for maintaining its credibility lay with the local coordinator, whose performance served as an important indicator of program quality. The coordinators all started with good credibility because they entered the scene soon after site selection (while overall program credibility was high), because they had Eichler's strong support, and because the hiring process was skillfully handled (Eichler selected them from a national pool and then had them interviewed and formally chosen by a committee of locally respected individuals).

As the program unfolded, the coordinators' ability to build and sustain their credibility (independent of their relationship with Eichler) depended on how well they cultivated and maintained strong relationships with members of the support community, particularly members of the LISC Local Advisory Committee. This, in turn, was shaped by several factors.

The first was how well the local program sponsor and host positioned them vis-à-vis the private sector. Initial positioning was strong in Palm Beach County, where the coordinator had strong ties to the Palm Beach Economic Council and the MacArthur Foundation, and in Little Rock, where the program was brought in under the aegis of major banks and the coordinator was based at the Chamber of Commerce. The Greater New Orleans Foundation provided the local coordinator with a weaker strategic position because although it is well regarded, it had comparatively weak ties to the corporate sector.

A second strong influence on the coordinator's relationship with the LISC Local Advisory Committee was the timing of the decision to hire a full-time local LISC staff person to manage the "core" LISC grant and loan pool.[4] New Orleans, recognized from the outset as the most difficult program site, had a number of existing CDCs with which LISC initially (but incorrectly) thought it would be able to work; hence, an experienced LISC program director was brought in quite early. The local coordinator there thus had less opportunity than his counterparts in the other sites to cultivate personal relationships with Local Advisory Committee members. The New Orleans LISC program officer recognized the strategic importance of his relationships with committee members and protected them accordingly. In contrast, the Little Rock program officer was hired quite late in the demonstration program, mainly because the local coordinator—who also recognized the importance of strong relationships with Local Advisory Committee members—took on some of the tasks normally performed by the LISC program officer to minimize the perceived need to hire an additional person. Palm Beach County represents the intermediate case. The program officer started work about halfway through the demonstration program and had no prior experience with LISC. She took the lead in working with the Local Advisory Committee, but not as forcefully as the seasoned program officer in New Orleans.[5]

Personal factors also came into play. The Little Rock coordinator devoted more, and more consistent, effort to his relationships with the support community than his counterparts, and his easy-going style fit well with the local culture. The Palm Beach County coordinator, whose direct, down-to-business (a.k.a. "northeastern big city") style matched the local culture less well, became progressively more focused on dealing with real estate production delays and their causes; other tasks received less attention as a result. Ironically, in New Orleans, the good fit between the coordinator's style and local ways of doing business worked against him. He attached less importance than the other two coordinators to the strategic role of relationships with the Local Advisory Committee and did not wish to challenge the LISC program officer's actions in making himself the primary program contact with the support community.

In the long term, the credibility of the program and the coordinators hinged on their success in moving the program steadily forward; the key indicator relied on by supporters was demonstrable progress on CDC real estate projects. When implementation was on schedule, the local coordinator's credibility was high. When the Little Rock coordinator departed after two years, the program's production track record was strong, and both he and the program were highly regarded. Major delays in production for all the New Orleans CDCs and all but two of those in Palm Beach County ultimately undercut both coordinators' credibility significantly. As a result, both experienced a decline in their ability to shape the program's direction effectively.

Credibility Among Community Volunteers. The keys to the program's credibility among the CDC volunteers were the community organizers. LISC's national reputation and a favorable reaction among residents who met with Eichler during the site selection process both carried some weight in targeted neighborhoods, but many of the volunteers were not "sold" on reputation and wanted to see real actions before "buying in" to the program.

Organizers who had roots in the neighborhoods where they were assigned to work enjoyed the highest credibility initially. This seemed to contribute to their early success in gaining commitments from volunteers. The fact that all three of the Little Rock organizers had strong connections to the neighborhoods they were working in contributed significantly to that program's fast start "out of the blocks" in gaining volunteer commitment and organizing CDCs. However, almost all the organizers in the three sites established their credentials well during their initial efforts selecting neighborhoods, as they spent hundreds of hours in each neighborhood and asked questions and listened like few others had done before. This enhanced the credibility not only of the organizers, but of the entire program effort.

The fact that the program did a good job in recruiting a diverse staff also helped to build its credibility at the community level. In each site, at least two thirds of the organizers and at least half of the local development team staff were people of color. This was noted and appreciated in the target communities, most of which were predominantly African American.[6]

After the initial reception, however, organizers had to prove themselves with hard work and results. Their credibility (like the program's progress) fluctuated over time, but was clearly correlated with their ability to help the volunteers get their work done. Helping with administrative tasks (e.g., photocopying, reminding people about meetings and deadlines) was routinely valued because most of the CDCs had no staff support. In the long term, it meant keeping the group moving toward the interim objective: a completed real estate project. In Little Rock, where technical support was strong, organizers were valued for their ability to help move the development process forward. In the other sites, the organizers' credibility also depended on their adeptness in helping to deal with the difficult process issues that failure to meet important goals brought to the surface.

Overall, LISC and the demonstration enjoyed high credibility in all three sites initially. That credibility declined with program experience in two of the three sites. The fact that the CDCs' real estate project performance fell well below expectations was the driving force behind this decline. The program had held out the promise of spawning community-based development capacity that would be capable of producing housing (and, later, other tangible improvements). Lack of housing was a problem in and of itself for many supporters, and without it even those funders whose priority was building community capacity could see little evidence that their goals had been met.[7] The loss of credibility was most evident in Palm Beach County, where prominent members of the support community were discouraged and disappointed about the program's failure to intervene when the coalition started experiencing difficulties, even after senior LISC officials promised to do so on several occasions.[8]

Lessons. The obvious lesson here is that no program can sustain real credibility for very long if it cannot show meaningful progress toward its objectives. But program experience also offers some guidance about how to promote program credibility. Getting an initiative off to a good start through the careful selection of local partners (e.g., host and sponsors in the LISC demonstration) is very much worth the effort it takes. Selection criteria should include the prospective partner's commitment to the intervention's goals and the means it will use to attain them; the degree to which that partner is well regarded locally, especially among those the intervention most needs to influence; and the partner's ability to be helpful. That said, a good start is just that: a start. It can carry the program only so far and must be followed up in a steady and consistent manner.

Designing interventions with specific steps and milestones—as this initiative did throughout the first half of the program—is an effective tool to build and sustain credibility over time; each milestone, when met, provides yet another assurance that the program is "on track" and in good hands. Establishing this type of track record also provides a hedge against risk; when (not if) something does go wrong, the credibility and goodwill established by a string of successes buys the program time to recover. If possible, building in a focus on more than one activity or indicator of progress (in this case, not only real estate development but also either organizational development or non-real-estate neighborhood improvements) is also useful in maintaining credibility and managing risk.

A third lesson is that the way a community development program deals with issues of color and gender affects its credibility. This lesson goes well beyond the short-term benefits gained from assembling a diverse staff. Managing these issues well enables the program to gain greater insight into the issues and dynamics in and among community groups in neighborhoods of color. Equally important, it signals the initiative's respect for targeted communities and its understanding of the importance of these issues to the community.

A final point is that credibility and the ability to effectively influence local actions will be affected by the congruency of programmatic efforts. For example, including women and people of color in meaningful roles in the demonstration program modeled the behavior and outcomes the initiative sought to promote: greater influence for historically excluded groups and respectful working relationships among people of diverse backgrounds.

■ Confidence

The confidence participants have in the program is strongly related to its credibility. Credibility has a strong reputational component, is directly related to the reality and perceptions of "hard results" (i.e., performance on program promises), and is most relevant to program partners and staff. Confidence has a stronger personal psychological element and is most relevant for the targeted population (the confidence they have in their own efficacy and the potential efficacy of program initiatives). The LISC demonstration devoted considerable attention to building volunteers' confidence in their collective ability to effect change in their neighborhoods. Confidence had a major effect on the volunteers' level of commitment, and on their proclivity to assume responsibility and control.

During the early months of the program, the confidence of volunteers increased substantially—bolstered both by the attention they received (from organizers, coordinators, and the support community) and by their early successes (e.g., in organizing CDCs and town hall meetings). Confidence fostered participation, as confident board members were more likely to attend meetings regularly, volunteer to do tasks or join committees, and actively recruit neighbors to participate in CDC-sponsored activities. CDCs that suffered early setbacks, such as those few whose town meetings were poorly attended, received prompt attention from their organizers, who both encouraged the volunteers and challenged them to do better.

Like program credibility, the self-confidence of many volunteers and supporters' confidence in them declined in New Orleans and Palm Beach County as production fell well behind schedule and hence became the focus of attention. Among volunteers, this loss of confidence contributed to a decline in participation that, in turn, made it more difficult for the CDCs to get things done. In more than half of the CDCs in these sites, the number of consistently active volunteers dropped to one or two.

In Little Rock, sustained progress on the CDCs' real estate projects, including accomplishments that met or even exceeded expectations, helped keep volunteer confidence high. This contributed to the volunteers' high level of participation in their CDC activities and their acceptance of the challenge to organize a coalition. It also contributed to the fact that the coalition agreed to serve as the lead agency in preparing Little Rock's application to the federal Empowerment Zone and Enterprise Community program. Despite subsequent setbacks for the coalition, the individual CDCs, their presidents, and a broad base of volunteers continued to be engaged; participation at CDC meetings remained high and the presidents' council continued to meet.[9]

Lessons. Many of the activities that help build program credibility among volunteers also help increase their level of confidence, but additional steps are also needed. A series of activities—each ambitious enough to require residents to "stretch" in some way, but modest enough to be achievable (in this program, e.g., holding town meetings, identifying a project, obtaining predevelopment funding)—that are visible and can be celebrated is an effective way to build volunteer confidence over time.

As with building credibility, it is useful to have different types of milestones being worked toward simultaneously, particularly when one objective is to engage volunteers with diverse interests, skills, and inclinations. There is no doubt that this makes program management more difficult and consumes program resources (e.g., organizer's time spent helping volunteers work on nondevelopment activities). On the other hand, it also helps to manage the risks of focusing all effort and attention on one type of milestone (e.g., real estate production).

The confidence of the volunteers needs to be monitored on an ongoing basis. In the demonstration program, this was comparatively straightforward, because the organizers—generally astute observers and good listeners—had a great deal of contact with the volunteers and developed good relationships with them. Contact in groups, although useful, is not enough. Working closely with individuals to develop their confidence is often required. Of particular importance is having regular contact to prepare key individuals, such as CDC presidents and committee chairs, for group meetings whose success is itself part of confidence building. This type of one-on-one work is also critical in developing leadership ability. Open and frequent conversations among volunteers can also be useful, especially when the program is going well. In Little Rock, for example, regular meetings of the CDC presidents helped sustain confidence at a high level; the CDCs could both learn of the successes of others and be acknowledged by their peers for their own accomplishments. This suggests the usefulness of building into the program design opportunities for volunteers (individually and collectively) to be self-reflective and discuss their personal and collective concerns and misgivings (e.g., annual retreats).

Occasional ebbs in volunteer confidence are likely to be unavoidable, especially among participants who may have experienced many setbacks and disappointments in the past; the program must be flexible enough to deal with these, and staff must be prepared for them and trained how to respond. Ironically, volunteers commonly "let down" and hesitated after a big accomplishment—reticent to take on a new challenge at which they might fail and thereby jeopardize their newly won victory. In these instances, the tactics employed in the demonstration program worked well: reinforcing the importance and value of what has been accomplished, reminding volunteers of the importance of their goals and the commitment they have made to them for the community, express-ing confidence that they can reach the next milestone, and assuring them that they will get the help they need. In cases where confidence hits a serious low, it may be necessary to devise interim, short-term (confidence-boosting) objectives that will be relatively easier to achieve. When low confidence is the product of poor outcomes because volunteers have not met their commitments, this issue must be addressed quickly, and head on.

■ Competence

Competence encompasses the technical, financial, and organizational skills of program staff, volunteers, and members of the support community. It is central to the creation of community development capacity: the ability of volunteers, program staff, and the support community to deliver visible improvements to engaged neighborhoods. Viewed from the perspective of Temkin and Rohe (1998), competence is essential to the new organization's ability to assume its role as an effective part of the neighborhood's institutional infrastructure. Through its influence on the delivery of results, competence also has a significant impact on volunteer confidence, program credibility, and, hence, commitment.

The demonstration raises two major issues surrounding the cultivation of competence. First, the program's inability to identify local sources of technical support for CDC development projects (e.g., to persistently coach and encourage volunteers through the detailed work involved with real estate development) in two of the three sites, and its slowness in putting effective alternatives in place, meant that the technical competence of volunteers to engage effectively in real estate projects was neither as broad nor as deep as it needed to be. Second, volunteers did not gain the degree of skills and experience in organizational matters (e.g., hiring and overseeing staff) that they needed as they moved into progressively more demanding activities. At least three factors contributed to this second shortcoming: the pressure to push the development projects forward, the organizers' inability to provide the degree of support necessary, and the fact that the program design paid insufficient attention to building these skills.[10]

Clearly, good program design and implementation must ensure that core competencies are developed vis-à-vis those activities and roles that will have the greatest impact on outcomes, that is, the short list of things that the intervention must do well for everything else to work. In this LISC demonstration program, these were real estate production and organizational development. Program experience offers three lessons about building local competency.

Lessons. The first lesson is that external efforts to "jump-start" community development in new places cannot assume the presence of adequate technical-assistance capacity locally. Rather, technical support should be part of the program design. Early, realistic assessment of local capabilities—in this case, to help CDCs deliver real estate products and cultivate their governance and decision-making skills—is important. In cities where community development has become well established, the availability of good local technical assistance developed over time in response to the demand for it (Vidal 1992). A similar (but probably more rapid) process is likely to be needed in new localities. Initiatives intended to "parachute in" community development from the outside should be prepared to cultivate and use locally available technical-assistance resources where they exist (to build local support capability), but should also be prepared from the outset to supplement what is available locally, as needed—perhaps for a period of several years. The contrast between Little Rock, where the program found ample technical support, and the other two sites illustrates the costs of not ensuring that solid technical assistance is in place: unmet timetables, diminished community confidence and engagement, and reduced program credibility.

A second lesson is that competencies needed by program participants are likely to change over time as circumstances, challenges, and opportunities change. Some changes can and should be anticipated in a good program design, but programs that are charting new waters cannot anticipate everything. Hence the wisdom of periodically taking stock of whether new types of technical support are needed, particularly when program participants (volunteers, funders, and staff) are about to move into new activities or roles. For example, as production increased in scale in Little Rock, the need for stronger project-management skills evolved and was addressed. In contrast, none of the sites adequately prepared volunteers to hire, oversee, and set policy for staff, either in their individual CDCs or in the context of a coalition.[11]

The final lesson is that capabilities should be more broadly considered than was the case in this demonstration. Beyond technical real estate competency, volunteers needed management and organizational skills; they acquired some, but in most instances needed more than they were given an opportunity to learn. Being more explicit about the leadership and organizational skills that volunteers required to succeed would have focused the program on the need for stronger volunteer preparation. It would also have provided a ready means to focus the attention of supporters on domains in which the CDCs were making progress (even when development projects were moving slowly), thereby deepening their comprehension of the varied skills CDCs require and the variety of ways in which they can contribute to the efforts of neighborhood volunteers.

■ Comfort

Program participants (staff, volunteers, members of the support community) lacking prior connection begin to relate effectively to one another when they have had some positive, shared experiences. Viewed in social capital terms, such experiences are vehicles through which participants identify and confirm their mutual interests and build relationships based on trust. Program experience illustrates clearly that fostering bonds among members of relatively homogeneous groups, in the context of shared objectives, is eminently doable, whereas fostering ties that bridge well-established social barriers is fraught with difficulty.

Comfort and trust take time to develop and are necessarily based on positive experience. Because the LISC demonstration program was relatively short and enjoyed mixed success, its track record in raising participants' comfort level with one another is necessarily mixed as well. Program supporters, particularly members of the LISC Local Advisory Committees, were already fairly comfortable with one another when the program began; most already knew one another, either professionally or through shared participation in civic activities. Moving from that foundation to comfort and trust around the specifics of the program required little explicit effort. The few exceptions to this pattern involved either individuals with very different perspectives (e.g., the African American County Commissioner in Palm Beach County, who agreed to serve for a time on the LISC Local Advisory Committee but remained suspicious of the program) or with different powerful relationships (e.g., major contributors with whom some other members of the committee were sometimes reluctant to disagree openly).

Volunteers' comfort with other members of their CDC boards developed fairly quickly in most cases. Some participants already knew (or knew of) one another, and, unlike the Local Advisory Committees, they met often and had to cooperate actively to complete a demanding series of tasks. Differences in board members' ability to master technical aspects of real estate development well enough to complete their share of the work and to participate confidently in making group decisions was an issue for some CDCs. As discussed in Chapter 5, CDCs faced this issue most directly after they received their predevelopment funding, and they handled the issue differently; participation sometimes declined as a result, but uneven understanding of (or interest in) real estate rather than lack of comfort or trust was the reason.

The program also made good progress in fostering comfort among volunteers from different CDCs. Volunteers knew from the outset that they all were part of the same program and hence had a common vocabulary and a shared set of experiences even when they did not know one another. This gave them a shared understanding of community development and laid a strong foundation for bonds among them. Local coordinators emphasized this as the time neared for the CDCs to decide whether to join together in a coalition, and a variety of site-specific events (e.g., the collective action taken by the Palm Beach County CDCs to press for changes in county policy, the site visits by members of the Mon Valley Initiative) reinforced the theme. The fact that the CDCs in all three sites favored the coalition approach suggests that this aspect of the program was working well.

On the other hand, neither supporters nor resident volunteers developed real comfort in dealing with each other across lines of class and race. Local coordinators linked volunteers to selected members of the LISC Local Advisory Committees when they were preparing to seek project funding from LISC and from local banks. These meetings were widely viewed as successful. The committee members who saw volunteers do a trial presentation of their project were typically impressed with (and often surprised by) the presentation's quality and the ability of the volunteers to respond well to questions; these committee members often became advocates for the projects they had visited. Volunteers valued both the substantive help and the respect they received and were obviously pleased when their funding applications were approved. These contacts were a good beginning when they happened (not all CDCs got to the point of seeking construction financing), but they were only a beginning, and there was little follow-up. In the absence of established relationships, funders in Palm Beach County and New Orleans were uncomfortable with the task of holding the CDCs accountable for what they had pledged to do; those in Palm Beach County and Little Rock did not have confidence in volunteers' ability to accept criticism, and hence were not comfortable voicing their reservations about the candidates that volunteers favored to direct their coalition organizations.

Lessons. Trust and comfort lay the foundation on which program participants with varied backgrounds and interests can come together to solve problems and resolve disagreements. They facilitate many types of transactions and encounters, but are especially important when program participants must address substantive issues that become entangled—as they commonly do—with larger, more charged issues: race, class, and power. The racially mixed CDC boards discussed in Chapter 5 all experienced difficulties of this kind. Even more difficult were issues that arose in different forms at all three sites during and after transition: Differences of opinion about the performance standards the CDCs had to meet to merit funding, about the relative importance of race as a criterion in selecting an executive director for the new coalitions, and about the role funders should play (if any) in the decision making of groups of volunteers all became entangled with strong feelings about race, control, and class- and culture-based differences in priorities and standards that few were comfortable talking about.

It is not realistic to expect an intervention of short duration to fundamentally change the character of divisive issues and of local conversations (or lack thereof) about them. Any externally stimulated program, to be effective, has to learn how to operate in the context of those divisive issues by helping groups develop sensitivity to others and break down negative stereotypes. Change in people's comfort level will surely be slow, but even helping a small number of well-regarded people—whites as well as blacks—and coaching them to exercise leadership on these issues can yield important payoffs.

Good communication (discussed later) can help foster comfort and trust, but program-related activities also need to be part of the program package. Actions provide participants with an opportunity to verify that others behave in ways that are consistent with the views and values they express in conversation—an essential element of trust. Actions that produce successful outcomes—for example, a strong CDC presentation to a bank officer who decides to extend a loan—are especially helpful. They help build participants' confidence in one another's capabilities and generate a reservoir of goodwill on which participants can draw to get through disagreements and miscommunications.

■ Critique

Critique involves program participants reflecting on their experience, their contributions, and the contributions of others; engaging in conversation with others on these reflections; and working with others to improve performance. In a complex, dynamic effort with multiple stakeholders, ongoing critique that incorporates the viewpoints of all major stakeholders can make a major contribution to long-run success. Critique is particularly valuable in strengthening community development capacity.[12]

In the LISC demonstration, shared critiques were mostly informal and ad hoc. The only formal vehicles for critique were the LISC Local Advisory Committees and the local coordinators' strategy committees. Both of these were underutilized for this purpose. The LISC Committees' critiques were focused on real estate performance measures, such as production, and did not consider a broader range of managerial, organizational, and system issues. The strategy committees, as originally envisioned, included critique as one of their possible functions, but were ineffectively organized.

This is not to say that participants did not reflect on what they were doing; they did. Local coordinators met weekly with their organizers to assess progress, talk about what was and was not working, and plan next steps. Coordinators talked as needed with Eichler and met with him during his site visits, to consider the bigger picture. Eichler and Manson had ongoing conversations about how the program was progressing and engaged in similar conversations with key funders in the sites.

However, the program did not set aside times in which groups of participants could reflect analytically on the program's progress and the relation of their own activities to that progress. Conversations that discussed participants' concerns and dissatisfactions, or that helped supporters understand the complexities and subtleties of the work, were typically small and involved a comparatively small number of players. As a result, opportunities for increasing participants' comprehension of community development, their comfort with one another, and hence their willingness and ability to reflect critically were lost.

Lessons. Establishing a program culture in which critiques are encouraged requires not only openness and trust, but also collective recognition that everybody will be better off in the long run if critiques are forthcoming. Initial successes that build confidence and engender respect for participants' capabilities and goodwill can help to break down the tendency of individuals and organizations to be defensive, and thus can foster an environment favorable for active critique.

Informal or small conversations that begin to introduce and legitimate critique are likely to be very helpful, but formal institutional vehicles can also be used once some groundwork has been laid. The LISC Local Advisory Committees and strategy committees could have been used more effectively to monitor and critique performance. Program experience suggests that the establishment of clear, fair standards and the tracking of performance by representatives of credible organizations (such as LISC) are valued by both funders and volunteers who are committed to making a real difference. It gives them clear direction, and when they know they are being watched they are more likely to perform better.

Standards, monitoring, and critiques should seek to focus attention on the full range of desired program outcomes. For example, the LISC demonstration program could have established performance measures for organizational and leadership development in targeted neighborhoods, for example, the number of active board members, the number with growing skills and competence in different aspects of development, nondevelopment activities the CDCs were sponsoring, and the boards' success in building working committees that integrated members from different ethnic, social, and community interest groups on CDCs. The development team monitored such matters internally as part of their ongoing internal process of critique, but the program did not do as much as it might have done to broaden those conversations.

Key Aspects of Program Strategy and Implementation

Devising and implementing strategies for complex initiatives is always a challenge and is to some degree idiosyncratic in each site. Nevertheless, the demonstration program suggests five broadly applicable principles.

■ Communication

A necessary precondition for comprehension, confidence, and constructive critiques is ongoing communication between and among program participants. One program goal was to establish many new conversations about community development, including those between LISC and potential funders in the support community; in low-income neighborhoods, among residents and between residents and program staff and members of the support community; and among members of the support community on the LISC advisory and strategy committees.

Many of the early program activities, such as CDC organizing and meetings, social and media activities, town meetings, real estate development project work, and strategy committee meetings, were designed, in part, to provide vehicles for communication. In all three sites, many different parties were engaged in new conversations about community development objectives, strategies, activities, and outcomes. Both the depth and breadth of communication appeared to increase in the early stages of implementation in all three sites.

Some conversations included the larger community, including the media, members of the public sector, and the general public. This was particularly true in New Orleans, where media coverage of the LISC demonstration highlighted how the program represented a new and better way of undertaking community development in the city.

Communication seemed to inspire commitment, and the flow of information and knowledge among participants enhanced community development capacity. For example, a relatively high percentage of volunteers waited to make their decision about whether to join a CDC until after they had had conversations with other prospective members and been to a meeting—illustrating the link between conversation and commitment. The support communities' comprehension of community development also increased through communications with staff and volunteers. The contrast between the character of interviews with members of the support community early in the program and the sophistication evident at the end of the program was striking.

Like some of the intermediate outcomes, communication growth abated after the CDCs were organized and focused in on their real estate projects. Communication became more problematic after real estate projects fell behind schedule in two sites and the development teams disbanded; opportunities for communications that would expand general comprehension declined.

Lower participation at CDC meetings in New Orleans and Palm Beach County meant there were fewer opportunities to put important issues on the table and to engage in collective discourse and problem solving. Compared with the first year of the program, LISC Local Advisory Committee communications focused less on increasing comprehension of community development and more on real estate project activity.

With less participation and communication in the context of "regular" program activities, management and implementation became more problematic. Staff had to meet with participants more and more on an individual basis to get their input and support. This had obvious efficiency costs, but more importantly made it more difficult for participants to overcome biases, increase their comfort working with one another, view issues from different perspectives, and formulate consensus. Staff increasingly served as intermediaries, which was incongruent with program objectives.

Lessons. Ongoing opportunities for communication among different participants in program activities are especially important for an initiative that relies on a consensus model and that seeks to promote local ownership and control. Staff have a delicate balance to maintain; they need to cultivate strong relationships with individual participants and coach them when necessary, but avoid adopting the role of an intermediary who "relieves" program participants of the responsibility to learn how to work together and resolve their differences. Staff can, however, act as *communication strategists,* doing the behind-the-scenes work needed to create opportunities for participants to communicate effectively.

Formal institutions are natural vehicles to promote communication. However, program experience suggests that the existence of these institutions does not ensure that meaningful conversations will transpire; they are likely to require encouragement and facilitation. Given the desire (and, in this case, the pressure) to demonstrate progress, formal organization meetings (e.g., CDC board meetings and LISC Local Advisory Committee meetings in this case) are likely to focus on day-to-day business if left to their own devices. Program staff and, in time, local leaders will likely have to be proactive (e.g., in setting agendas, selecting and motivating participants, or inviting guests) in encouraging conversations that are more thought-provoking and that help build ties and comfort among people with different points of view.

Special effort is needed to stimulate constructive communications on important and difficult issues that may be uncomfortable for participants to raise on their own, but need to be raised. Leading examples of these types of issues in the community development field are race and control of development activities. One of the objectives of the LISC demonstration could have been to engage participants in conversations about these issues at opportune points. There is no doubt that this would have been difficult—but perhaps not more difficult than dealing with the problems that arose when these issues began to affect the course of the program without getting an open and forthright airing.

■ Consistency

Consistency of vision and purpose, both among key intervention agents (i.e., national LISC staff including Michael Eichler, development teams, and COI in this demonstration) and over time, helps to give a complex intervention coherence. It allows the agents to coordinate their activities effectively and provides a touchstone to which local participants can return in times of difficulty or confusion. It has two dimensions: operational (e.g., whether different agents are working well together, or who assumes responsibility when there are problems) and strategic (i.e., whether there is agreement among agents on program objectives and approach).

In the early stages of the demonstration, the lead agents' roles were clearly defined. Richard Manson and Michael Eichler, both LISC employees, worked closely together on site assessment and fund-raising. Once the program was under way, the program differentiated clearly between the role of the development team (focused on organizing CDCs and serving as a resource for the fledgling groups) and that of LISC core program staff (focused on working with the Local Advisory Committee and managing the grant and loan pool). This separation made good use of the agents' skills and was designed to be attractive to both potential funders (thought to be most interested in establishing standards and accountability for funds disbursement) and community residents (striving for independent control of development in their communities). It also provided a way to let participants know that whereas LISC was making a long-term commitment to work in the locality, the services of the development team were of limited duration.

Participants varied considerably in their understanding of the relation between the two elements, both within each site and across sites. Occasionally, this created some confusion. For example, when the CDCs were ready to apply to LISC for funding, some volunteers wondered, "how can the same organization be both a coach/advocate and take a 'tough banker' approach with us?"[13] Throughout most of the program's duration, however, problems of this kind were quite modest.

As discussed in Chapter 7, as the transition date neared and housing production fell below expectations, the program's consistency of vision and priorities, and hence of strategy, failed. In the program design, by transition the CDCs were supposed to have demonstrated their capacity to engage in development with a level of proficiency that met national standards and, working either through a coalition or on their own, to be capable of managing development in their communities. The reality was that most of the CDCs in Palm Beach County and New Orleans had made less progress than expected, and the two coalition organizations that had been created had pronounced vulnerabilities. When it became clear that the goals of housing development and volunteer control would not both be met within the program time frame, the main program agents had to choose—and they chose differently.

The most obvious signal of this was Michael Eichler's departure from LISC and his establishment of an independent organization, the Consensus Organizing Institute (COI); this changed the relation between the two components of the program. This created some confusion and disagreement, especially in Palm Beach County, around the issue of who was responsible for providing guidance and organizational development for the new coalitions. However, Eichler's departure was arguably more a symptom of more fundamental differences of opinion than the cause. The more fundamental problem was that Eichler and the development teams gave higher priority to maximizing community control of development through the solidarity and efficiency provided by a coalition, whereas LISC gave higher priority to real estate development. As it became evident that the LISC and COI teams did not share the same priorities and were not coordinating their efforts, both lost credibility in the eyes of program participants.

Lessons. The clear lesson is the importance of community development partners (including intervention agents) sharing the same vision and priorities. Differences in priorities lead to confusion and a paucity of clear direction. Up front, and throughout implementation, partners should discuss priorities and the implications of priorities for programmatic actions. Disagreements must be worked out before they lead to confusion and dysfunctional actions "on the ground"; any necessary changes or accommodations should be made with sensitivity for their implications for program participants, particularly those who are most vulnerable—in this case, the volunteers.

■ Congruence

Congruence refers to how well a community development initiative's activities and tactics and the actions and words of key actors match overall program strategy and objectives. Program experience suggests that congruence has a strong influence on overall program credibility and on the credibility of individual participants. It also has an influence, both directly and through its effect on credibility, on sustaining the commitment of volunteers and members of the support community. In short, those implementing the intervention serve the program well if they model the behavior they seek to stimulate.

The demonstration offers many good illustrations of program tactics and staff actions that mirror the program's strategy and objectives. Most of the program's energy and effort were expended at the neighborhood level, consistently signaling that building the community's capacity to assume responsibility was a central objective. The use of a diverse group of local people, drawn from both the neighborhoods and the support community, to interview development team candidates was congruent with the strategy of building bridges between the support community and the neighborhoods. The racial diversity of the team itself mirrored the program's insistence that CDC boards be inclusive. The community organizers met with hundreds of neighborhood residents; they were respectful, were interested, cared about the neighborhood, and worked hard—exactly what they wanted from volunteers—and their consistency in these respects won the volunteers' respect and confidence.

Incongruence, when it occurred, took its toll, costing the program and its staff credibility and respect. This happened primarily around the issue of control during the transition—when the extent and character of the transfer of control was uppermost in everyone's mind. One of the main objectives of the LISC demonstration was to increase local control of community development. Congruence thus required that the program effort respect local individuals and organizations and their potential to exert control of community development effectively.

Among the CDC volunteers, the most unfortunate example of incongruence was the ill-fated effort of the Palm Beach County local coordinator to influence the selection of the coalition executive director (see transition discussion in Chapter 7). She felt that the individual the volunteers ultimately selected was a very poor choice, and she proved to be correct. Unfortunately, her handling of the situation alienated volunteers when they were most sensitive about issues of control. The program had emphasized that when the development team departed, it would be their turn to assume control; but when it was time for them to take over, they perceived that they were not being trusted. The fact that they were having real difficulty making the decision, and that their difficulty was evident, only made the situation more volatile.

A very different type of incongruence affected volunteers in New Orleans. When transition was imminent and none of the CDCs had even started a development, LISC encouraged each CDC that had project plans to consider undertaking a joint venture with a private developer. Bacatown CDC chose this option and won praise when its first project broke ground, but the private developer was clearly in the lead role. LISC also brought in new technical staff to work with the CDCs. Some volunteers found their heavy reliance on outside technical consultants and the decisions of "removed actors," including bankers and government bureaucrats who controlled the funding, was disempowering, especially when compared to the empowering nature of the front-end organizing efforts. This contributed to the decline of confidence among volunteers, and some CDC members started to feel that the objective of community control might be beyond their reach.

In the support community, the perception in New Orleans and Palm Beach County was that LISC's actions after the problems experienced at transition were not sensitive to the preferences and abilities of local actors. In New Orleans, some prominent members of the Local Advisory Committee felt that their input and preferences were not adequately considered by the local program director when he made decisions. Many felt that by adding LISC staff (the technical consultants, accountable to the program director), the program became further from, not closer to, local control. In Palm Beach County, key members of the Local Advisory Committee felt that LISC (and, to a lesser extent, COI) had the responsibility and expertise to assist the local program in dealing with the weaknesses of the coalition—but that neither lived up to their promises.

These issues were obviously controversial, and participants on both sides of each issue had strongly held, albeit divergent, views. What is clear, however, is that program actions that are seen locally as being incongruent with core program principles and objectives can be costly—even if the actions taken have an apparently strong rationale.

Lessons. Using the details of program design and implementation to model and reinforce an intervention's goals, strategies, and values requires considerable care and attention to detail. It also pays dividends, because it demonstrates in a tangible way that the espoused goals and values are actually taken seriously. This can be especially important for members of poor communities, who commonly feel they have had ample experience with people and programs that say one thing and do another.

Perceptions matter. To be effective, program staff have to be sensitive to how their actions, management styles, and personalities are perceived, and how those perceptions influence participants' view of the program. More broadly, however, the experience of this program suggests that there are real advantages to identifying program staff who not only have the skills to do the job, but also have values consistent with those of the initiative. The everyday behavior and attitudes of such staff members model and reinforce their formal efforts.

Congruence, like other aspects of maintaining a sensitivity to process issues, requires explicit thought and attention. It can make activities or decisions take longer to complete if the quality of the outcome is to be maintained. In addition, there are some indications from the LISC demonstration that pursuing congruence consistently and well raises participants' expectations, making instances of incongruence more disturbing.

■ Counterbalancing

Counterbalancing refers to the central challenge of using outside intervention and directed support to build local capacity and inspire the taking of local control and responsibility. As illustrated by program experience in all three sites, managing this tension was no easy task, particularly at transition. This challenge

and that of adapting implementation to the local context (discussed subsequently) are both vital to effecting commitment, capacity, and control. Much
of their effect is indirect through their influence on the other key program
attributes.

Because the need for counterbalancing permeates all aspects of externally
driven interventions, it assumes many forms that may appear to be discrete issues
to program participants but that are, in fact, part and parcel of the same core task.
The LISC demonstration program provides some excellent examples. These
include the following:

1. The tension between the need to set and keep firm deadlines to motivate participants and to build confidence (when the deadlines are met) and the importance of
   being flexible and adaptable to local circumstances so as to encourage local
   comfort and nurture local capacity.

For example, in Palm Beach County, moving back the transition date by 18
months resulted in a loss of momentum. A shorter "window" might have better
inspired staff and volunteer action and commitment—the reason the development team targeted a 12-month transition date instead. In Little Rock, holding
to the two-year transition date at first seemed efficacious, but later it became
evident that volunteers lacked capacity to effectively oversee a coalition organization. In New Orleans, linking extension with the requirement of specific
results (i.e., groundbreakings by all the CDCs) seemed to inspire staff and
volunteers and retain support community commitment.

2. The tension between process and product. The goal is to produce valued, visible
   intermediate outcomes quickly enough to inspire commitment and gain credibility, but to go slowly enough to build competence and confidence among a broad
   cross section of participants.

For example, although there were benefits to linking the program's extension
to real estate production targets in New Orleans, that link also forced production
to proceed at a pace too fast to foster volunteer learning. "Partnerships" between
CDCs and private developers (e.g., in Bacatown) were undertaken in which the
private developer did most of the work, with only marginal volunteer involvement and correspondingly little learning. A contrasting example with a similar
capacity-building outcome was South Little Rock CDC, in which volunteers
wanted to do everything themselves—but they made very limited progress on
real estate production, and capacity gains came slowly. Argenta CDC in North
Little Rock may be the best example of effective pacing of production; volunteers strongly influenced the staging and timing of development efforts, starting

with single-family housing rehab projects (one at a time) and moving on to progressively more complex development activities.

3. The tension between providing the strong leadership and guidance by staff, technical assistance providers, and the support community that volunteers need to help structure unfamiliar work and decisions, while still allowing those volunteers the opportunity to assume greater control and increase their capacity by learning, making meaningful choices, and sometimes delaying program activities or making mistakes.

In Palm Beach County, the local coordinator ran into difficulties and lost credibility when her recommendations for executive director of the coalition were perceived as being too controlling and not supporting volunteer decision making. A similar situation was evident in New Orleans, in the relationship between the LISC program director and some members of the support community; some members of the support community felt strongly that the program director was too controlling and did not value either local opinions or local control.

In contrast, the discussion of building relations with the support community highlighted how a lack of direction and clear statements of priorities and opinions can be detrimental. For example, in Little Rock we observed that the coalition executive director—and through him, the coalition itself—could have benefited from more coaching and mentorship from members of the support community or local LISC program director.

4. The tension between requiring that inexperienced groups meet national community development standards and adapting expectations to local circumstances. Meeting national standards can help build program credibility and foster commitment from the support community. However, it can also be debilitating to local staff and volunteers if standards are difficult to achieve given the local context—a subject we will take up in greater detail in the next section.

Lessons. A key lesson—albeit a vexing one—from the LISC demonstration program is that there are no easy methods or pat solutions to the challenge of counterbalancing the tensions in externally driven community development interventions. To be successful, program sponsors and staff must recognize the existence of these built-in tensions and design strategies and tactics to manage them effectively. This includes training or coaching for staff and key participants (e.g., the community organizers and lead local sponsors in the LISC demonstration), so they will recognize the tensions as they appear in different guises and have tactics at hand to manage them.

Such effective management requires taking steps to see that these tensions (or the management of them) are not perceived as incongruencies by participants, that is, as a program failure to do what it has promised or to act in accordance with its espoused principles. In some instances, it may be advantageous to make the tensions transparent to program participants, so that learning about them becomes part of the ongoing effort to strengthen participants' comprehension. For example, program experience suggests the value of having CDC volunteers be cognizant of the trade-offs between going too fast or too slow in pacing their development project activity. Many groups in the demonstration had a tendency to do either one or the other without considering the trade-offs involved, but (as the discussion in Chapter 5 noted) CDCs whose presidents were sensitive to the need to balance process and product did so effectively, moving their projects forward without sacrificing learning among board members unfamiliar with real estate. In other cases, making the tensions transparent may be inappropriate or even detrimental to the program. For example, if participants know that a deadline has been set as part of a strategy to get them to work harder, the deadline may lose much of its motivational power and possibly provoke resentment.

One way to monitor whether counterbalancing efforts are going well is to routinely assess participants' level of engagement and effort and the extent to which they show initiative. If these are high, they provide a signal that participants have the commitment, the capacity, and the perceived degree of control they need both to maintain their commitment and to move program efforts forward. If not, remedial action is likely to be needed.

One pitfall that is important to avoid is allowing either community residents or members of the support community to become too dependent on staff—particularly staff whose role is intended to change or diminish over time (like the development teams in the LISC demonstration). Staff, especially those close to major program transitions, should pay particular attention to the implications of their actions for the willingness and ability of community members and their supporters to assume greater control and responsibility; modify their roles in the direction of acting as coaches (rather than managers or outside experts); and focus on the legacy they will leave (i.e., how their actions will affect the potential for sustained local activity). In some cases, it may also be beneficial to enlist members of the support community as mentors to community residents and/or local community development staff if they are prepared, or have been prepared, to assume this role.

Program experience suggests that, in general, setting firm, clear deadlines that are challenging but realistic is an effective way to maintain program momentum, energy, and credibility. If changes in major targets or deadlines are clearly justified and in the program's long-term best interest, that justification and the new deadlines and expectations should be clearly articulated to all program participants. Linking changes in deadlines to specific required actions (which must be realistic to avoid any additional disappointment, but which also must be challenging) can be beneficial. To be so, they must involve open, honest communication among the parties and be generally perceived as fair.

■ Context

Adapting program implementation to the local context is (like counterbalancing inherent tensions) a pervasive issue, although it is commonly a low-profile issue because it is generally taken as "given" and most local participants give it little explicit thought. The LISC demonstration program's site-selection process took into account some aspects of the local context, especially the low level of previous community development activity, the supportiveness of the private sector, and the level of interest among prospective volunteers. All were critical ingredients of program success.

Other differences in local context, and the program's varying success in adjusting program implementation to reflect them, also affected program outcomes. Important among them were the local political climate and culture (including its compatibility with the program's objectives and style), the history of local community development (especially any negative experiences that would shape local attitudes and reactions), the degree of social and political openness to people and ideas from outside the area, and attitudes toward race (as shaped by the local history and current state of race relations).

In many respects, the LISC program's adaptations to local context were effective and paid dividends. The choices made in staffing the development teams are a good example, and the implications of these early choices rippled throughout the program's implementation. The three demonstration sites fall along a spectrum in terms of the program's response to local circumstances.

In New Orleans, the importance attached to having a team that was predominantly African American and that was comprised of local people was a direct response to local conditions and attitudes. In Little Rock, the local coordinator limited his search for organizers to the locality, convinced that people from the local African American community would be best able to win the confidence of prospective volunteers. He himself was not from the city, but his genial, easygoing style matched the local culture well, and he readily formed good relationships with a wide range of people in the support community. In Palm Beach County, where in-migration is common and the neighborhoods from which the program would select its target communities are somewhat more racially diverse, the fact that the coordinator and one organizer were both white and from Pennsylvania posed little difficulty either in the support community or in the targeted neighborhoods.

A second important example is the program's site-specific strategies for dealing with the local public sector. In New Orleans, where local government operated on a patronage system (and was viewed by many as being corrupt) when the program began, the program avoided all contact with city government. Instead, the local coordinator and LISC program director worked with a nonpartisan good-government group to lay the foundation for a neighborhood-oriented city housing policy proposal that was later adopted by a new, honest mayor. Several of the CDCs have received funding from the new administration. Little Rock presented the opposite situation: a highly capable local government that initiated efforts to attract LISC and its consensus organizing program to the city. The program had a good relationship with the city from the outset, and the CDCs' applications for city funding for their projects gave the city an opportunity to develop its own capacity by learning about how to deal with nonprofit developers.

Palm Beach County presented a more mixed picture and was therefore the most difficult site for which to develop and implement a strategy. The two County Commissioners elected from the neighborhoods the program was likely to serve were of different persuasions: one inclined to support the program, the other not. Key members of the county's civil service also were mixed in their attitudes toward the program, and those in line positions were very comfortable doing things as they were used to doing them. The program's strategy was a mixed one as a result. Staff cultivated supporters informally and helped the CDC volunteers confront the nonsupporters when that was their only recourse.

On the other hand, the program adapted too little and too late to an unanticipated aspect of local context: the paucity of appropriate technical assistance providers in two of the three sites. As the analysis in Chapters 5 and 7 makes clear, and the discussion of competence earlier in this chapter highlights, this was a costly shortcoming.

To the extent that one of the program's objectives is to change how things are done locally, adapting the program to the local context becomes more difficult and problematic. The most pointed example of this in the LISC demonstration program was New Orleans. The cultural norms of taking things at their own pace and of tolerating patronage and other forms of corruption (i.e., "the Big Easy") made it an especially difficult environment to implement a program structured around very different norms: timely progress on projects and programs that met national standards for integrity and technical quality. The fact that "outsiders" were the carriers of the new standards only made the situation more difficult.

Operationally, the local culture in New Orleans made it very difficult to make progress on community development projects, because many of the local participants involved had low expectations and standards, especially regarding timeliness and meeting commitments. Efforts by the local LISC program director to press for these things were viewed by some volunteers and members of the support community as being arrogant and disrespectful of local culture and capability. Only a few members of the LISC Local Advisory Committee kept in focus the fact that corporate interest in sponsoring LISC hinged on LISC's perceived ability to change "how things are done" in the city. On the other hand, the local coordinator (a New Orleanian) sympathized with many aspects of local culture and did not press the CDC volunteers hard enough to stimulate a critical mass of real estate production; he, too, suffered some loss of credibility and respect as a result.

Lessons. Awareness of, and sensitivity to, the local context is integral to effective implementation. This requires that external agents be attentive to gauging how outside actions are being perceived locally. Ideally, a program would have access to assessments of local perceptions and reactions from several points of view, particularly assessments by individuals who are good observers of the often subtle dynamics surrounding issues of race and power. Program experience suggests that the assessments of knowledgeable, well-placed local actors who are committed to the program's success can be a valuable addition to the assessments made by program staff. Needed adjustments to program implementation so that local reactions are not detrimental to program progress and outcomes can then be made.

All externally driven community development interventions are efforts to stimulate local change. The more ambitious (or comprehensive) the initiative's objectives, the greater the likelihood that some aspects of the program will be at odds with some local organizations, norms, and cherished points of view. Hence, being sensitive to the local context typically cannot mean leaving it unchallenged and unchanged. What the New Orleans experience emphasizes again, however, is the importance of setting realistic goals and managing expectations so as to sustain commitment to the program and changes it seeks to effect. Dividing the program into stages, and using each stage to take stock of how well the change process is going and whether any changes in implementation tactics or strategy are called for, appears to work well. Conversely, failure to take corrective steps in a timely fashion clearly carries costs.

The premise of the consensus organizing approach, which the experience of the LISC demonstration indicates has considerable promise, is that the challenges an intervention poses to aspects of the local situation need not be direct and are most effective when influential local actors share responsibility with external program agents in pressing for change. The broad strategy applied in this case appears to be quite generalizable. Start with local partners who are committed to program goals (even though they may not fully understand their implications, including the fact that reaching those goals may require them and their organizations to change how they go about their work). Work consistently to deepen the understanding and commitment of local partners, and encourage and help them to take the lead in pressing for and facilitating change. Simultaneously, use both formal and informal opportunities to gradually broaden the circle of critically placed individuals who understand the program and are sympathetic with its objectives. Use confrontation only when necessary and pick your battles carefully, keeping in mind that local program participants may have to continue to work with those who seek to thwart the program's progress.

Broader Reflections

The LISC demonstration program was an attempt by an external agent with local supporters to create new bonding and bridging social capital in the service of community development. The results are instructive: encouraging in their demonstration that meaningful change is possible in a variety of places, sobering in their reminder of the difficulty of the task.

Consensus Organizing

The success of this program in facilitating citizen participation, identifying and supporting a cadre of new neighborhood leaders, enhancing the development capacity of the residents of targeted low-income neighborhoods, and deepening the commitment and understanding of external supporters to community development argues that consensus organizing as an approach to creating institutional infrastructure to serve low-income communities shows considerable promise. This view is reinforced by LISC's perspective that the two sites added later to the demonstration, which enjoyed the benefits of the lessons learned in these three early sites, developed more smoothly and with better results.

As the analysis presented in the discussion of major activities seeks to make clear, the key to its success—and arguably to similarly complex community development efforts of other types—does not reside in any one activity or program element, but in their combination. For this reason, a strategic approach, in which each major activity is structured to promote multiple objectives as part of a long-term process, is central to the program's effectiveness. Structuring major program activities so that they serve as both means and ends, although quite difficult to execute consistently, appears to pay considerable dividends; each activity entails a process that has intermediate value but also advances the program toward its long-term goals.

Community Participation

The consensus organizing approach of building a new community organization that cuts across existing networks and cultivates new leadership is clearly more difficult than more conventional approaches, which typically work through existing networks and with existing leaders. However, both theory and practice suggest that it is likely to be a stronger strategy in many, perhaps most, places where effective community-based organizations are not already at work.

From a conceptual perspective, conventional approaches can, just as this program did, fill structural holes in target communities by creating new organizations—either CDCs or other types of entities, such as service delivery collaboratives or neighborhood governance structures—that become part of the community's institutional infrastructure. The apparent superiority of this approach lies in its going beyond conventional approaches to emphasize making community networks more dense by bridging across existing patterns of association in the neighborhood.

Program experience suggests two ways in which the approach pays off in practice. By building organizations with broad representation, in a programmatic context in which community members will set the agenda for their own organization, a program increases the likelihood that the activities the group chooses will enjoy broad support in the community and will actually be in the community's interest. In addition, this kind of cross-cutting representation gives the new organizations credibility, both in the neighborhoods and in the external political and funding environment.

These advantages are not always deeply appreciated by community residents, particularly in the early phases of the intervention. Representativeness as a value has appeal, but at a practical level building and running a new organization is more difficult if participants are new to one another and have divergent points of view. The temptation among people who are working hard to get things done is to fall back on familiar ties. This is especially true when new volunteers are needed, either because the group's work requires more people-power (e.g., to work on new activities) or because participants who leave the group for some reason must be replaced.

Absent conscious effort to maintain breadth of participation—meaning not just getting greater numbers of people, but getting diverse segments of the community—representativeness tends to diminish over time. Sustained outside organizing activity appears to be needed, at least until the new group's leadership and organizational culture have internalized the importance of broad representativeness as an organizational asset. One or more group experiences that demonstrate the value of diversity in tangible, pragmatic ways that help further the group's goals are likely to be needed before this occurs.

One way to encourage breadth and continuity of participation is to structure the intervention so that it offers both a range of opportunities for participation (e.g., options that require different types of time commitments or different kinds of activities) and reasons for commitment. Not all potential participants are motivated by the same incentives. Hence, a diversity of incentives and opportunities appears to be advantageous, as long as the underlying goal, such as revitalization of a neighborhood, is shared.

The prospect of control and self-determination is a motivator of resident participation—crucial to attracting and retaining high-quality participants, to their successful development, and to motivating their continued engagement. Actual progress in increasing residents' ability to give effective voice to their priorities and concerns and to decide what will be done in response is also a major source of legitimacy for both the community group and the initiative itself.

Community development interventions other than those with an organizing focus might take a useful lesson from this program's emphasis on preparing residents of target communities carefully and well before thrusting them into positions of responsibility or settings in which prospective external supporters will be likely to assess their capabilities. Many members of the support community will have had limited interaction with residents of targeted communities and are likely to have preconceptions about their abilities that can undermine the quality of any collaborative effort. The experience of the LISC demonstration suggests very strongly that early meetings or joint activities in which residents play their roles confidently and well begin to break down such preconceptions and enable residents to enter shared deliberations or activities in a much stronger and respected position. In many instances, residents will need substantive assistance, coaching, or practice in order to get their engagement with prospective partners off to a good start. Depending on the circumstances, they may need assistance preparing for other development activities as well.

Private Sector Participation

The demonstration program's emphasis on identifying sites where the program would enjoy significant private sector participation sets it apart from many other community development approaches. Exceptions include other programs sponsored by LISC; those sponsored by the Enterprise Foundation, which also seeks local, private financial support in its sites; and the EZ/EC program, which encourages (but does not require) direct private participation in the program but does look for private sector investments that complement publicly funded ones.

This program made more demands on members of the local private sector than is typical even in other LISC sites, especially at the time when those other sites first began working with LISC. This was particularly the case as transition approached and during the transition process. Members of the LISC Local Advisory Committees were told during the initial fund-raising process that contributors would have the ability to guarantee that the local program would be of high quality because they, through their participation on the committee, would be responsible for approving both policies and expenditures. For many private contributors, this assurance was an important part of their decision to support LISC.

When the program encountered difficulties and the time came to exercise their ability to help uphold program quality, however, many committee members were reluctant to use their authority, especially in face-to-face meetings with CDC board presidents or other community volunteers. In retrospect, it is clear that they expected program staff to ensure that they would never actually have to play that role. Some program sponsors and staff were quite disappointed about this, but—again in retrospect—it is clear that committee members had not been adequately prepared for that role, in part because they did not want it. As a result, they intervened actively too little and too late.

That said, private sector participants in these sites have learned a great deal about what successful community development requires—learning perhaps best capsulized in the candid but rueful reflection of a senior executive, "We should have known this was going to be hard, but we didn't." The fact that they remain committed to the program is a very positive sign, and not only because it bodes well for the likelihood that they will continue to be knowledgeable supporters of community development in their localities. It confirms prior experience, both in the Mon Valley and in the LISC demonstration program when the CDCs were developing their funding proposals, that under better circumstances, the private sector would be willing to be more actively engaged in community development. Specifically, interventions pursuing this goal would need to clear what was expected, cultivate an understanding of why it is important, and prepare private sector participants to play their new roles well. This willingness (both in these sites and in other places) is likely to be contingent, however, on program interventions that seek to engage them in maintaining a "results" orientation— one of LISC's strongest appeals to its private sector constituents.

Public Sector Participation

There appears to be room to improve on the program's approach toward the public sector. During the site selection process there was some tendency among senior program staff to presume that if some form of community development was not already ongoing in a site, then at least some parts of the public sector—which has a responsibility to serve and improve low-income neighbor-hoods—were not doing their job well, and would therefore be potential sources of opposition to the initiative and its objectives. As a result, the program design (i.e., "the model") had no formal public sector component. Instead, it built in the pessimistic view by deferring direct engagement with the public sector until comparatively late in the intervention and assigning the main responsi-bility for garnering public sector support to the newly formed CDCs (with the program's behind-the-scenes help) at the point when they needed public sector help, for example, subsidy dollars to complete their project financing. Given the many factors that shape local public policy, this presumption seems unduly pessimistic.

In these sites, the public sector was more receptive to supporting and working with the program than senior staff initially assumed and than the program design reflected. Local coordinators identified and made good use of this support where it existed, but might have been better prepared and positioned to do so if early recognizance and program strategizing had been more open in thinking about possible roles for members of the public sector. Some CCIs, particularly those that have a long-term systems reform agenda, and the EZ/EC program, are more positive about the possibility of engaging the local (and sometimes state) public sectors and have explicit strategies for doing so. It will be interesting to observe how well these efforts fare.

In sum, the LISC consensus organizing demonstration program provided a limited test of the feasibility and usefulness of trying to build bridging and bonding forms of new social capital that would (1) strengthen the institutional infrastructure serving selected low-income communities and (2) lay the foundation for expanded activity in the future. Its experience suggests that social capital construction is difficult and slow, but it is possible, and that good program design and implementation can cultivate an environment in which it will grow more rapidly than it would if no one intervened. Its experience also confirms both the special difficulty of building bridging capital across lines of race and class and the obvious fact that that is where the need is greatest.

Notes

1. More specifically, as described in Chapter 2, the program had clear objectives at two levels. At the community level, they included increased (1) comprehension of community development; (2) participation in community development; (3) organization of community development; and ultimately (4) control of community development. The program also sought increased support—financial, technical, and political—of community development from private and public sectors (what we call here *the support community*). The most difficult and potentially controversial of these objectives, and the one around which there emerged disagreement with regard to priority and timing, was community control of development.

2. Systems knowledge stands in contrast to technical knowledge of real estate and organizational development, which are discussed under competence later in this chapter.

3. The experience with coalitions also reflects in part, as suggested in the transition discussion, that supporters' expectations of volunteers were too high given the degree of preparation they received.

4. Richard Manson, based in New York City, managed the core program in each site until the volume of activity increased to a level that justified a full-time, on-site staff person. Once hired, that person had primary responsibility for working with the Local Advisory Committee (as in all LISC sites).

5. The program officer in New Orleans had previously worked for LISC in Washington, D.C.

6. The program did draw some criticism for being an "outside, white" organization because senior LISC management— the two managers of this program and their superiors in New York— were white. We heard these criticisms mainly from African American skeptics or opponents of the program who also had other reasons for holding a negative point of view, but it seems likely that the criticism nevertheless struck a responsive chord with some members of the African American community.

7. This feeling was strongly reinforced for some by the general decline in volunteer participation in many of the CDCs whose projects had stalled, although some funders were better informed about this than others.

8. For some, particularly in New Orleans, the perception that the program's actions had become incongruent with its stated goals also raised credibility issues. This is discussed in the section on congruence later in the chapter.

9. This somewhat anomalous result reflects both the fact that the CDCs had the capacity to continue to do useful work to which residents were committed and the fact that the delegates to the coalition were, in general, not among the leadership in their individual CDCs.

10. These issues are discussed in depth, with relevant examples, in Chapters 5 and 7, but the issue of the organizers' inability to provide needed organizational development assistance is especially difficult to assess. The organizers were selected on the basis of their potential to become good organizers, not real estate development or organizational development technical assistance providers. Some additional training on organizational issues would have prepared them to do more, but would not have solved the basic problem; some organizers had their hands full simply trying to support project progress and help the volunteers meet the day-to-day demands of keeping the organization going. The more sophisticated skills needed by the volunteers, especially to succeed in managing their coalition organizations, called for more extensive and specialized help than the organizers could reasonably have been expected to provide.

11. The sequencing of competency development needs explicit attention. Participants are being asked to learn many new things. The time and energy they can devote to the program are finite, and varied things they need to learn are best conveyed in different ways (e.g., working one-on-one with a coach, extended interaction with a group at a series of meetings or an intensive retreat, formal instruction, or simply through practice). Identifying capabilities that must have top priority (as noted in Lesson One) is key and may involve difficult trade-offs.

12. Critique is also an aspect of program management. It is treated here as an intermediate outcome because it is unrealistic to expect participants to be comfortable critiquing the program (and implicitly one another) during the initial stages of an intervention effort. Positive program progress and the cultivation of trusting relationships among varied program participants must first lay the groundwork.

13. This was most commonly an issue for volunteers from CDCs in Palm Beach County and New Orleans that received some technical help from LISC in preparing their project materials because the technical team envisioned in the program design was not in place.

References

Bates, Timothy. 1994. "An Analysis of Korean-Immigrant-Owned Small-Business Start-Ups to African-American and Non-Minority-Owned Firms." *Urban Affairs Quarterly* 30(2):227-48.

Bratt, Rachel G., Langley Keyes, Alex Schwartz, and Avis Vidal. 1994. *Confronting the Management Challenge: Housing Management in the Nonprofit Sector.* New York: Community Development Research Center, New School for Social Research.

Briggs, Xavier de Souza. 1998. Forthcoming. "Brown Kids in White Suburbs: Housing Mobility and the Multiple Faces of Social Capital." *Housing Policy Debate* 9:1.

Brown, Prudence. 1996. "Comprehensive Neighborhood-Based Initiatives." *Cityscape* 2(2):161-176.

Burt, Ronald. 1992. *Structural Holes: The Social Structure of Competition.* Cambridge, MA: Harvard University Press.

Chaskin, Robert J. and Mark L. Joseph. 1995. *The Ford Foundation's Neighborhood and Family Initiative: Moving Toward Implementation: An Interim Report.* Chicago, IL: The Chapin Hall Center for Children at the University of Chicago.

Chaskin, Robert J., Selma Chipenda Dansokho, and Mark L. Joseph. 1997. *The Ford Foundation's Neighborhood and Family Initiative: The Challenge of Sustainability: An Interim Report.* Chicago, IL: The Chapin Hall Center for Children at the University of Chicago.

Coleman, J. 1988, Summer. "Social Capital in the Creation of Human Capital." *American Journal of Sociology* 94:S95-120.

Coleman, James. 1990. *Foundations of Social Theory.* Cambridge, MA: Harvard University Press.

Committee for Economic Development. 1995. *Rebuilding Inner-City Communities: A New Approach to the Nation's Urban Crisis.* New York: Author.

Connell, James P., Anne C. Kubisch, Lisbeth B. Schorr, and Carol H. Weiss, eds. 1995. *New Approaches to Evaluating Community Initiatives: Concepts, Methods, and Contexts.* Washington, DC: The Aspen Institute.

181

The Council for Community-Based Development. 1993. *Expanding Horizons III: A Research Report on Corporate and Foundation Grant Support of Community-Based Development.* Washington, DC: Author.

Foley, Michael and Bob Edwards. 1996. "The Paradox of Civil Society." *Journal of Democracy* 7(3):38-52.

Gittell, Marilyn, Kathe Newman, Janice Bockmeyer, and Robert Lindsay. 1996. "Expanding Civic Capacity: The Urban Empowerment Zones." Paper presented at the annual meeting of the Urban Affairs Association, March 14, New York.

Gittell, Ross. 1992. *Renewing Cities.* Princeton, NJ: Princeton University Press.

Gittell, Ross and Philip Thompson. 1996. "Inner-City Business Development and Entrepreneurship: New Frontiers for Policy and Research." Presented at the National Community Development Policy Analysis Network Conference, Brookings Institute, November 16, Washington, D.C.

Gittell, Ross, Avis C. Vidal, and Margaret Wilder. 1997. *The Role of Enterprise Zones in Community and Economic Development.* New York: Community Development Research Center, New School for Social Research.

Goetz, Edward G. 1993. *Shelter Burden: Local Politics and Progressive Housing Policy.* Philadelphia, PA: Temple University Press.

Granovetter, Mark. 1973. "The Strength of Weak Ties Hypothesis." *American Journal of Sociology* 78(6):1360-80.

Granovetter, Mark. 1974. *Getting a Job: A Study of Contacts and Careers.* Cambridge, MA: Harvard University Press.

Green, Roy E., ed. 1991. *Enterprise Zones: New Directions in Economic Development.* Newbury Park, CA: Sage.

Hebert, Scott, Kathleen Heintz, Chris Baron, Nancy Kay, and James E. Wallace. November, 1993. *Nonprofit Housing: Costs and Funding, Final Report, Volume I.* Washington, DC: U.S. Department of Housing and Urban Development.

Hornburg, Steven P. and Robert E. Lang. 1998. Forthcoming. "What Is Social Capital and Why Is It Important to Public Policy?" *Housing Policy Debate 9:*1.

Jacobs, Jane. 1961. *The Death and Life of American Cities.* New York: Vintage.

Jargowsky, Paul. 1996. *Poverty and Place.* New York: Russell Sage Foundation.

Johnson, James H., Jr. and Walter C. Farrell, Jr. 1997. "Snapshots of income inequality in metropolitan Los Angeles. (1997 November). *Looking Ahead, XIX* (2-3), 39.

Keyes, Langley, Alex Schwartz, Avis Vidal, and Rachel Bratt. 1996. "Networks and Nonprofits: Opportunities and Challenges in an Era of Federal Devolution." *Housing Policy Debate* 7(2):21-28.

Kubisch, Anne C., Prudence Brown, Robert Chaskin, Janice Hirota, Mark Joseph, Harold Richman, and Michelle Roberts. 1997. *Voices From the Field: Learning From the Early Work of Comprehensive Community Initiatives.* Washington, DC: The Aspen Institute.

Kubisch, Anne C., Carol H. Weiss, Lisbeth B. Schorr, and James P. Connell. 1995. "Introduction." Pp. 1-21 in *New Approaches to Evaluating Community Initiatives: Concepts, Methods, and Contexts,* edited by James P. Connell, Anne C. Kubisch, Lisbeth B. Schorr, and Carol H. Weiss. Washington, DC: The Aspen Institute.

Leiterman, Mindy and Joseph Stillman. 1993. *Building Community: A Report on Social Community Development Initiatives.* New York: Local Initiatives Support Corporation.

Local Initiatives Support Corporation. 1994. *1993 Annual Report.* New York: Author.

————. 1997. *1996 Annual Report.* New York: Author.

Marris, Peter and Martin Rein. 1973. *Conclusion of Dilemmas of Social Reform: Poverty and Community Action in the United States.* Chicago, IL: Aldive.

Mills, Edwin S. and Luan Sende Lubuele. June, 1997. "Inner Cities." *Journal of Economic Literature* 35:727-57.

National Congress for Community Economic Development. 1995. "Tying It All Together: The Comprehensive Achievements of Community-Based Development Organizations." Washington, DC: Author.

Oliver, Melvin L. and Thomas M. Shapiro. 1995. *Black Wealth, White Wealth.* New York: Routledge.

Perry, Stewart E. 1973. "Federal Support for CDCs: Some of the Issues and History of Community Control." Cambridge, MA: Center for Community Economic Development.

Powell, Walter. 1990. "Neither Market nor Hierarchy: Network Forms of Organization." In *Research in Organizational Behavior*, Vol. 12, pp. 295-336, edited by B. Straw and L. Cummings. Greenwich, CT: JAI.

Putnam, Robert D. 1993. *Making Democracy Work: Civic Traditions in Modern Italy.* Princeton, NJ: Princeton University Press.

Putnam, Robert D. 1995a. "Bowling Alone: America's Declining Social Capital." *Journal of Democracy* 6:65-78.

Putnam, Robert. 1995b. "Tuning In, Tuning Out: The Strange Disappearance of Civic America." The Ithiel de Sola Pool Lecture, American Political Science Association.

Putnam, Robert. 1996, Winter. "The Strange Disappearance of Work." *The American Prospect* 24:34-48.

Shiffman, Ronald and Susan Motley. 1989. *Comprehensive and Integrative Planning for Community Development.* New York: Community Development Research Center, New School for Social Research.

Temkin, Kenneth and William Rohe. 1998. Forthcoming. "Social Capital and Neighborhood Stability: An Empirical Investigation." forthcoming. *Housing Policy Debate* 9:1.

"United Way of America's Housing Initiatives Program" [United Way in Housing/Community Development, UnitedWay.org homepage, World Wide Web, December 1997].

U.S. Bureau of the Census. 1992. *Census of population and housing, 1990,* STF3A. Washington, DC: Author.

Vidal, Avis C. 1992. *Rebuilding Communities.* New York: Community Development Research Center, New School for Social Research.

Vidal, Avis C. 1995. "Reintegrating Disadvantaged Communities Into the Fabric of Urban Life: The Role of Community Development." *Housing Policy Debate* 6(1):169-230.

Vidal, Avis C. 1996. "CDCs as Agents of Neighborhood Change: The State of the Art." In *Revitalizing Urban Neighborhoods,* pp. 145-163, edited by Dennis Keating, Norman Krumholz, and Phil Star. Lawrence, KS: University Press of Kansas.

Vidal, Avis C. 1997. "Can Community Development Re-Invent Itself? The Challenges of Strengthening Neighborhoods in the 21st Century." *Journal of the American Planning Association* 63:429-438.

Vidal, Avis C., Arnold M. Howitt, and Kathleen P. Foster. 1986. *Stimulating Community Development: An Assessment of the Local Initiatives Support Corporation.* Cambridge, MA: State, Local and Intergovernmental Center, Kennedy School of Government, Harvard University.

Walker, Christopher. 1993. "Nonprofit Housing Development: Status, Trends, and Prospects." *Housing Policy Debate* 4(3): 369-414.

Warren, Rachael and Donald Warren. 1977. *The Neighborhood Organizer's Handbook.* South Bend, IN: University of Notre Dame Press.

Wilson, William Julius. 1987. *The Truly Disadvantaged: The Inner City, the Underclass, and Public Policy.* Chicago, IL: University of Chicago Press.

Wilson, William Julius. 1996. *When Work Disappears: The World of the New Urban Poor.* New York: Knopf.

Wright, David J., Richard P. Nathan, Michael J. Rich, and associates. 1996. *Building a Community Plan for Strategic Change: Findings From the First Round Assessment of the Empowerment Zone/Enterprise Community Initiative.* Albany, NY: Nelson A. Rockefeller Institute of Government, State University of New York.

Index

About the Authors

Ross Gittell is Associate Professor of Strategic Management and Public Policy at the University of New Hampshire's Whittemore School of Business and Economics. His areas of expertise and research include community development and state and local economic development. He is the author of *Renewing Cities* (1993), which compares community economic development efforts in four older industrial cities, and "Inner City Business Development and Entrepreneurship," (with Phil Thompson) in William Dickens and Ronald Ferguson (Eds.), *The Future of Community Development: A Social Science Synthesis* (forthcoming). He has also published in numerous academic journals, including *Economic Development Quarterly, Journal of Policy Analysis and Management, Regional Studies, National Civic Review, Journal of Entrepreneurial and Small Business Finance,* and *New England Economic Review.* His major current research and policy activities include community development research projects funded by the Rockefeller Foundation (evaluations of the Neighborhood Entrepreneur Program in NYC and the Foundation's Democracy Round Table initiative) and work with the state of New Hampshire and the New Hampshire Business Industry Association on economic development, income disparity, and fiscal issues in the state.

He is a member of the National Community Development Policy Analysis Network, New England Economic Study Group (at the Federal Reserve Bank of Boston), Urban Affairs Association, and Association for Public Policy

Analysis and Management. Prior to joining the faculty of the Whittemore School, he taught at Harvard University, where he was also a research fellow at the John F. Kennedy School of Government's Center for Business and Government and consultant for the Harvard Institute for International Development. In addition, he has taught at the New School for Social Research, where he was also senior associate at the Community Development Research Center. He received his PhD (in Public Policy) from Harvard University, an MBA from University of California at Berkeley, and an AB (Economics) from the University of Chicago.

Avis Vidal is Principal Research Associate in the Metropolitan Housing and Communities Center at the Urban Institute in Washington, D.C., where she is developing a program of research that will use a further elaboration of the constructs of social capital and community capacity to analyze the role of both community organizations and local and national institutions in shaping the well-being of low-income communities and their residents.

She served for 10 years as the founding Director of the Community Development Research Center at the New School for Social Research, where her research focused on the work of CDCs and led to such publications as *Rebuilding Communities: A National Study of Urban Community Development Corporations* and "Reintegrating Disadvantaged Communities Into the Fabric of Urban Life: The Role of Community Development." She has also served on the faculty at the Kennedy School of Government at Harvard University and as a member of the Legislative and Urban Policy Staff at the U.S. Department of Housing and Urban Development. She has published in major academic journals, including the *Journal of the American Planning Association, Urban Affairs Quarterly,* and *Housing Policy Debate.*

She is currently an Urban Land Institute Fellow. She is a member of the American Institute of Certified Planners, the Urban Affairs Association, and the Association of Public Policy Analysis and Management, and serves on the editorial boards of the *Journal of the American Planning Association* and the Center for Urban Policy Research Press at Rutgers University. She received her MCP and PhD in Urban Planning from Harvard University and her BA from the University of Chicago.